CHEMISTRY RESEARCH AND APPLICATIONS

PHOTOCHEMISTRY

NEW RESEARCH

CHEMISTRY RESEARCH AND APPLICATIONS

Additional books in this series can be found on Nova's website under the Series tab.

Additional e-books in this series can be found on Nova's website under the e-book tab.

CHEMISTRY RESEARCH AND APPLICATIONS

PHOTOCHEMISTRY

NEW RESEARCH

ABIGAIL B. SMITH
AND
NATHAN A. JOHNSON
EDITORS

New York

For permission to use material from this book please contact us:
Telephone 631-231-7269; Fax 631-231-8175
Web Site: http://www.novapublishers.com

Additional color graphics may be available in the e-book version of this book.

LIBRARY OF CONGRESS CATALOGING-IN-PUBLICATION DATA

Photochemistry : new research / [edited by] Abigail B. Smith and Nathan A. Johnson.
pages cm
Includes bibliographical references and index.
ISBN 978-1-62808-615-7 (soft cover)
1. Photochemistry. I. Smith, Abigail B. II. Johnson, Nathan A., 1962-
QD708.2.P465 2013
541'.35--dc23

2013028318

Published by Nova Science Publishers, Inc. † New York

CONTENTS

PREFACE

The study of chemical reactions that proceed with the absorption of light by atoms or molecules is known as photochemistry. Everyday examples include photosynthesis, the degradation of plastics, and the formation of vitamin D with sunlight. In this book, the authors discuss new research in photochemistry, including the photoprotective role of mycrosporine-like aminoacids; bicarbonate-enhanced photocatalytic degradation of phenol; conformational relaxation of excitons in conjugated polymers; and the quenching of flavin triplets by amino acids and hydrophilic carotenoids. (Imprint: Novinka)

The threat from ultraviolet (UV) radiation has been present on Earth before the development of any kind of life. The ozone layer is responsible for filtering most of the UV radiation, but some still can enter the biosphere and affect life. Moreover, due to ozone reductions, the level of UV radiation has increased in many parts of the world and it is likely to increase even further in the near future. Several evolutionary adaptations have allowed living organisms to cope with UV radiation. Specifically, the damaging effects of UV radiation can be diminished by several photoprotective mechanisms, such as accumulation of UV absorbing substances which act as sunscreens. Mycosporine-like amino acids (MAAs) are a family of substances occurring in a wide variety of marine and terrestrial organisms known to have a photoprotective role, among other functions. All these compounds are low-molecular-weight water-soluble molecules capable of absorbing UV radiation with absorption maxima ranging between 310 and 360 nm. The UV screening function of MAAs and related compounds can be inferred from theirphotophysical features, such as strong UV absorption, lack of fluorescence and high photo-stability.

In Chapter 1, we will summarize the biosynthesis and location of MAAs in different organisms, together with the isolation and structural characterization of MAAs. Also, we will describe the photophysical and photochemical properties of MAAs with an emphasis in the photoprotective role of these compounds.

In Chapter 2, sodium bicarbonate is able to inhibit the TiO2-assisted photocatalytic degradation of phenol in aqueous solution. However, in the presence of excess 2-propanol the transformation of phenol can be enhanced by bicarbonate. A likely explanation is that TiO2 under irradiation oxidises the carbonate and bicarbonate anions to the carbonate radical (CO3−•), which reacts selectively with phenol and does not react (or reacts only to a very limited extent) with 2-propanol. NaHCO3 would, therefore, shift reactive species from the transformation of 2-propanol to that of phenol, which could increase the degradation rate of phenol in the presence of bicarbonate. The enhancement effect by bicarbonate would be operational in the presence of organic matter in excess, and at bicarbonate levels that would be typical of waters with elevated alkalinity.

Conjugated polymers draw considerable interest due to their electroluminescence and photovoltaic properties, and are seen as cheap alternatives to inorganic based optoelectronic devices. The underlying physics behind the properties of these compounds is rather complex and may involve solvent relaxation, intra or inter chain energy transfer and conformational relaxation of the polymer backbone. Since the exciton is not delocalized over the entire chain but remains confined to just a few segments of the main backbone, the use of oligomers as model compounds enables us to disentangle several of the above mentioned photophysical processes. The problem of conformational relaxation in non polar solvents has been addressed by the authors (DOI: 10.1002/cphc.201200869) using a stochastic model based on a mechanochemical analog that includes the participation of the oligomer side chains in storing and dissipating the stresses induced by photoexcitation. The model was applied to a series of PPV (poly-phenylenevinylene) trimers to predict the relaxation rate constants over a wide range of solvent viscosities as well as its temperature dependence. By treating transport properties as local quantities, the model allows prediction of viscosity in the nano-volume surrounding the oligomer.

In Chapter 3 we propose to review in considerable detail the main features of our stochastic model extending the applications to new systems, in particular those which are predicted to relax in the sub pico-second timeframe such has PPV's without (or very small) side chains and/or P3HT's (3,4,4-trihexyl-2,2:5,2-terthiophene) poly-thiophenes which have a considerably smaller number of side chains then the previously reported systems.

Since non polar solvents such as chloroform are widely used to solubilize polymer films an extension of our viscosity model to polar solvents will also be addressed.

In Chapter 4 the photochemical behavior of the flavin chromophore is a matter of considerable interest, not only because the chromophore is an integral part of numerous flavoproteins, but also because riboflavin, or vitamin B_2, is an essential micronutrient and a component of light-sensitive beverages such as milk and beer. The photodegradation of riboflavin-containing beverages is attributed to the generation (following absorption of light and intersystem crossing) of triplet flavin, which sensitizes the formation of singlet oxygen, a highly toxic species. Since the flavin triplet is known to be quenched by suitable substrates (for example the amino acids tyrosine and tryptophan), sensitization of singlet oxygen can be blocked by the addition of an appropriate substrate. According to previous reports, tryptophan and tyrosine quench triplet flavin, but crocin (a naturally occurring hydrophilic carotenoid) does not; the quenching occurs through electron transfer from the quencher to the flavin triplet and the rate constant of quenching in aqueous buffer is rather large ($> 10^9$ M^{-1} s^{-1}). The present study was motivated by two considerations: Firstly, since carotenoids are known to be good electron donors and efficient triplet quenchers, the reported failure of crocin to quench the flavin triplet warrants a reinvestigation of the quenching kinetics; secondly, studies conducted over the last four decades have led to widely discordant values for the rate constants of the processes which govern the decay of triplet flavin in aqueous media, and there is clearly a need to identify the source of the discrepancies. The present study has revealed that crocin is nearly as good a quencher as tryptophan, and that quenching of flavin triplets by crocin and two other hydrophilic carotenoids proceeds through two parallel channels: electron transfer from the carotenoid to triplet flavin and energy transfer from triplet flavin to the carotenoid. For tracking the concentration of triplet flavin, detection of delayed fluorescence provides a far more convenient alternative (to laser flash photolysis) in general, and becomes particularly advantageous in systems containing carotenoids, where absorption by the carotenoid radical cation (produced through direct photoionization as well as electron transfer) overlaps that by triplet flavin. The practical advantages of using naturally occurring hydrophilic carotenoids (such as crocin and norbixin) for the photoprotection of riboflavin-containing beverages are also discussed.

In: Photochemistry: New Research
Editors: A. B. Smith et al, pp. 1-20
ISBN: 978-1-62808-615-7

Chapter 1

Photoprotective Role of Mycosporine-Like Amino Acids

Raúl Losantos and Diego Sampedro*
Departamento de Química, Centro de Investigación en Síntesis Química (CISQ), Universidad de La Rioja, Madre de Dios, Logroño, Spain

Abstract

The threat from ultraviolet (UV) radiation has been present on Earth before the development of any kind of life. The ozone layer is responsible for filtering most of the UV radiation, but some still can enter the biosphere and affect life. Moreover, due to ozone reductions, the level of UV radiation has increased in many parts of the world and it is likely to increase even further in the near future. Several evolutionary adaptations have allowed living organisms to cope with UV radiation. Specifically, the damaging effects of UV radiation can be diminished by several photoprotective mechanisms, such as accumulation of UV absorbing substances which act as sunscreens. Mycosporine-like amino acids (MAAs) are a family of substances occurring in a wide variety of marine and terrestrial organisms known to have a photoprotective role, among other functions. All these compounds are low-molecular-weight water-soluble molecules capable of absorbing UV radiation with absorption maxima ranging between 310 and 360 nm. The UV screening function of MAAs and related compounds can be inferred from their photophysical features, such as strong UV absorption, lack of fluorescence and high photo-stability.

In this contribution, we will summarize the biosynthesis and location of MAAs in different organisms, together with the isolation and structural

* E-mail address: diego.sampedro@unirioja.es (Corresponding Author)

characterization of MAAs. Also, we will describe the photophysical and photochemical properties of MAAs with an emphasis in the photoprotective role of these compounds.

1. Introduction

The importance of light for the development, preservation and sustenance of life has been recognized for a long time. Even ancient civilizations worshipped the sun in different forms as the ultimate energy source that allows life to be present in our planet. All the chemical transformations that take place in the biota are directly or indirectly induced by the sun and the light energy that provides. For instance, photosynthesis is the source of food for plants and most algae and sunlight is supposed to have played a relevant role in the origin of life on earth. Also, light is a relevant stimulus that can trigger many different biological processes: visual perception,[1] photosynthesis[2] and photomovement responses.[3]

However, it should be noted that not all the effects of light on living organisms are beneficial. In fact, high-energy portions of the solar spectrum can cause a significant damage to living cells. Only a minor proportion of the sunlight that reaches the biosphere corresponds with ultraviolet (UV) radiation characterized by a short wavelength (<380nm) but large potential danger to organisms. UV light can be absorbed by many different types of (bio)molecules and it could induce subsequent photochemical reactions that may ultimately lead to cell damage and death. Proteins, nucleic acids and diverse metabolites can efficiently absorb UV light and trigger diverse photoprocesses. Thus, most living organisms have developed several photoprotective mechanisms through evolution to allow them to cope with risks associated with UV radiation. Some metabolites, like melanin, try to protect the cells by absorbing light and thus avoiding damage in more sensitive molecules, such as DNA.[4] In case of DNA absorption, the cells can also repair or substitute the damaged portions.[5] Indirect protection can be provided by pigments such as carotenoids which can quench the reactive oxygen species generated by the radiation.[6] A very common strategy to avoid UV-damage is the presence or synthesis under UV-stress of secondary metabolites that efficiently absorb potentially harmful light. These compounds should absorb light (mostly UV radiation) and dissipate the energy as heat preventing damage to the cells.

Several types of metabolites are known to protect living organisms against harmful light. Scytonemin is a very stable pigment found in cyanobacteria that effectively screen the cells from UV and blue radiation.[7] Melanins are a group of compounds that share a red-brown color, water insolubility and broad UV absorption.[8] This type of compounds confers photostability to different living organisms, from fungi to superior animals. Carotenoids indirectly protect cells by quenching photosensitized products and trapping free radicals rather than by direct absorption of UV light.[9]

Mycosporines (also known as mycosporine-like amino acids or MAAs) constitute a singular class of photoprotective compounds. These secondary metabolites are very water-soluble, low-molecular weight, colorless products that share a similar chemical structure with small variations, although MAAs may also appear as glycosylated derivatives. Their UV spectra show a single, strong absorption band in the 250-350 nm region. This fact, together with their high photostability and lack of luminescence contributes to the efficiency of these compounds as sunscreens in many types of organisms.

In the present chapter we will discuss some of the features that make MAAs relevant secondary metabolites. First, we will briefly introduce the location and biological significance of MAAs. In the next sections, we will describe the isolation and structures of MAAs and related compounds. Then, we will show the diverse roles that these compounds can play in different organisms with an emphasis on the photoprotective role.

2. Location and Biosynthesis

Mycosporine-like amino acids are present in many organisms in different parts of our planet. This type of UV-absorbing metabolites[10] appears widespread among the fungal classes *Zygomycetes*, [11]*Deuteromycetes*, [12]*Basidiomycetes*,[11] *Bacilliarophyta* [13] and *Aphyllophorales* [13] but it is absent in *Agricales*,[14] and in the protozoan investigated.[13] Besides, these metabolites are found on the surfaces of apples exposed to solar radiation[15] and recently they have been also found and isolated in a broad array of marine species,[16] ranging from algae,[10] bacteria,[17]symbionts like dinoflagellate algae associated with scleractinian corals,[18] vertebrates,[13] embryos of shrimp,[19] lens tissues of fish[20] and in the external mucus that covers samples of *Fungiidae*.[21] However, they have not been detected in members of the genus *Ctenophora* and in certain holothurians.[13]

This abundance in many different types of organisms suggests that this type of compounds should have some properties that make them useful for the life of the organisms, more than a residual metabolite. Analyzing the distribution around the planet of the species which feature these compounds, it is found that they are present in all latitudes, ranging from template and tropical climates to the frozen Antarctica's oceans.[10, 13, 16] Thus, these compounds appear in all type of ecosystems and organisms, but are more relevant in aquatic species, fungi, bacteria, algae and plants; and especially in the external parts of cells or epithelial cells in more complex organisms.

The biogenesis of mycosporine-like amino acids is still unclear, but possible hypotheses[22] includes a β-diketone which origins in the shikimate pathway.[23] However, these metabolic routes are only viable in fungi, bacteria and plants.[24] Thus, animals should gain these compounds through diet. Data obtained[22a] from the incorporation of ^{14}C-labeled acetic acid, glyceric acid and 3-dehydroquinic acid (3-DHQ) are in agreement with a pathway in which the 3-DHQ, that is a metabolite of the shikimate pathway, acts as an intermediate or precursor of the mycosporines. Thus, 3-DHQ could act as intermediate in the biosynthesis of MAAS as shown in Scheme 1.

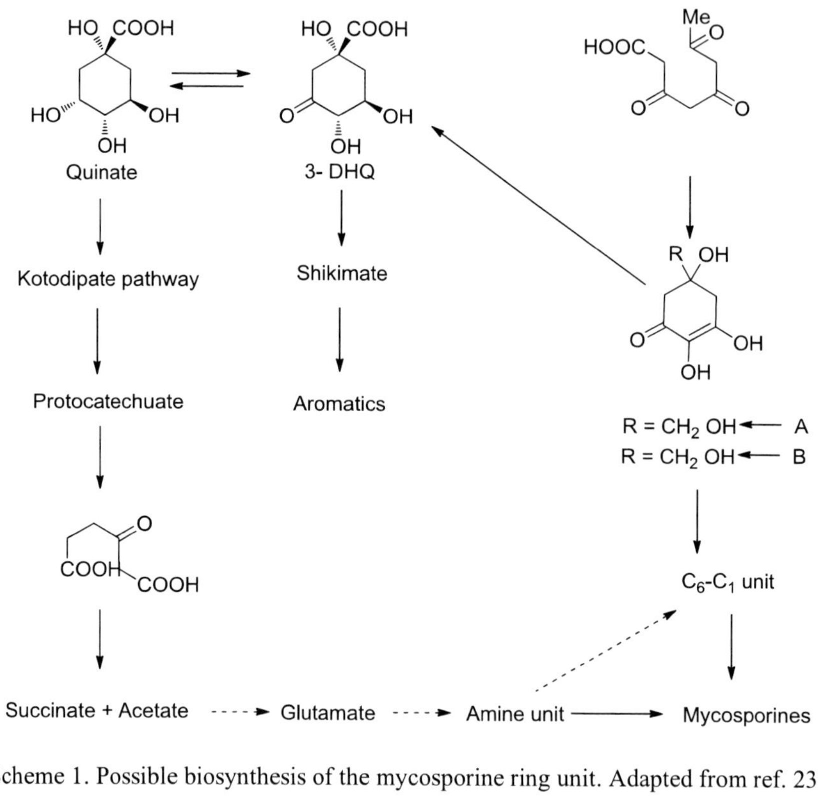

Scheme 1. Possible biosynthesis of the mycosporine ring unit. Adapted from ref. 23.

In contrast, more recent results[25] contradict the extended assumption of the biosynthesis of MAAs from the shikimate pathway. According to some evidences, the most probably bio-synthetic route could be the pentose phosphate pathway, but the primary steps of a shikimate-like route couldn't be discarded. Also, the use of ^{14}C-serine and ^{14}C-glycine,[26] allowed the detection of the specific incorporation into the corresponding side chains of MAA, demonstrating that these free amino acids are direct precursors and that mycosporine-glycine is necessary for the synthesis of bi-substituted imino-mycosporines, like shinorine.

The shikimate pathway is essential for the intermediate metabolism and it is functionally linked to photosynthesis. Thus, it is reasonable to think about a relationship between the presence of UV light and the activation of the biosynthesis or major accumulation of these compounds. Recent experiments[27] conclude that the incubation of copepod species and their concentration of MAAs are in relation with light conditions, finding a high concentration in the species incubated in presence of UV-radiation.

3. Isolation and Identification

Sampling procedures depends on the polarity and chemical stability of the source and the composition. Commonly, the sample consisting in algae, bacteria or marine organisms are treated with HPLC grade solvents for extracting the organic compounds present. Then, it's possible to follow three different procedures:[28]

- For phytoplankton, lyophilized or filtered samples are cut in centrifuge tubes and methanol is added. After centrifugation the decanted extracts are collected and filtered.
- Samples are soaked with small volumes of water overnight and in darkness, and then the first procedure is applied.
- If required, more aggressive conditions and solvents can be applied. Specifically, 25% aqueous methanol at 45°C through several cycles can be used until exhaustive extraction is achieved.

The obtained extracts for all procedures are then evaporated to dryness and re-dissolved in a mobile phase consisting of a pH = 3 solution of trifluoroacetic acid and 0,2%, ammonium hydroxide.

After recovery from natural sources, the methods used for the purification of these compounds are usually diverse chromatographic techniques. Gel permeation,[18] ion exchange resins,[29] norite A,[30] and preparative TLC on silica gel[31] have been used in different combinations to purify the samples. As other drugs, the characterization and structural elucidation is achieved by chemical degradation, spectroscopic techniques and X-ray analysis. As a general-purpose process, purification on Dowex 50W and carbon column followed by semi-preparative reverse-phase HPLC is usually a convenient method.

These general procedures of extraction and purification can be modified depending on the MAA or type of sample. Also, the mobile phase or the chromatographic column should be chosen according to the specific features of the sample and the metabolite.

4. Structure

As explained before, MAAs appear in a wide range of organisms and ecosystems. Interestingly, these compounds also present a very similar structure which contains a substituted cycle, specifically an aminocyclohexenone or an aminocyclohexenimine. Fungal metabolites that present a UV absorption band at 310-320 nm possess exclusively the aminocyclohexenone ring and are commonly

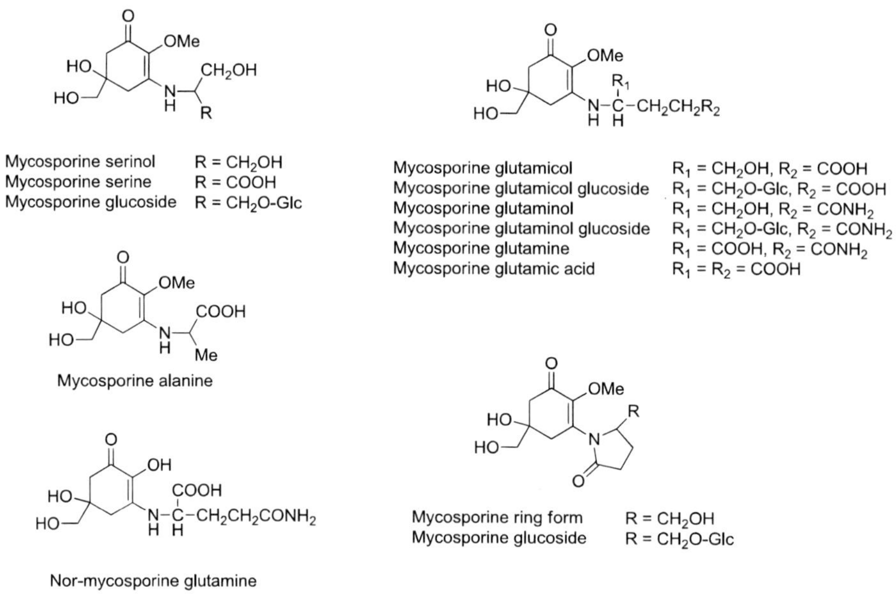

Figure 1. Fungi mycosporines structures.

known as mycosporines. In this type of compounds, two possible chromophores responsible for the UV absorption have been described (see Figure 1). Mycosporines have a methoxy group in C2 position and feature a band with a maximum at 310 nm while nor-mycosporines have an hydroxyl group in C2 position and a maximum at 320nm.[32] This basic structure can be subsequently modified in different points.

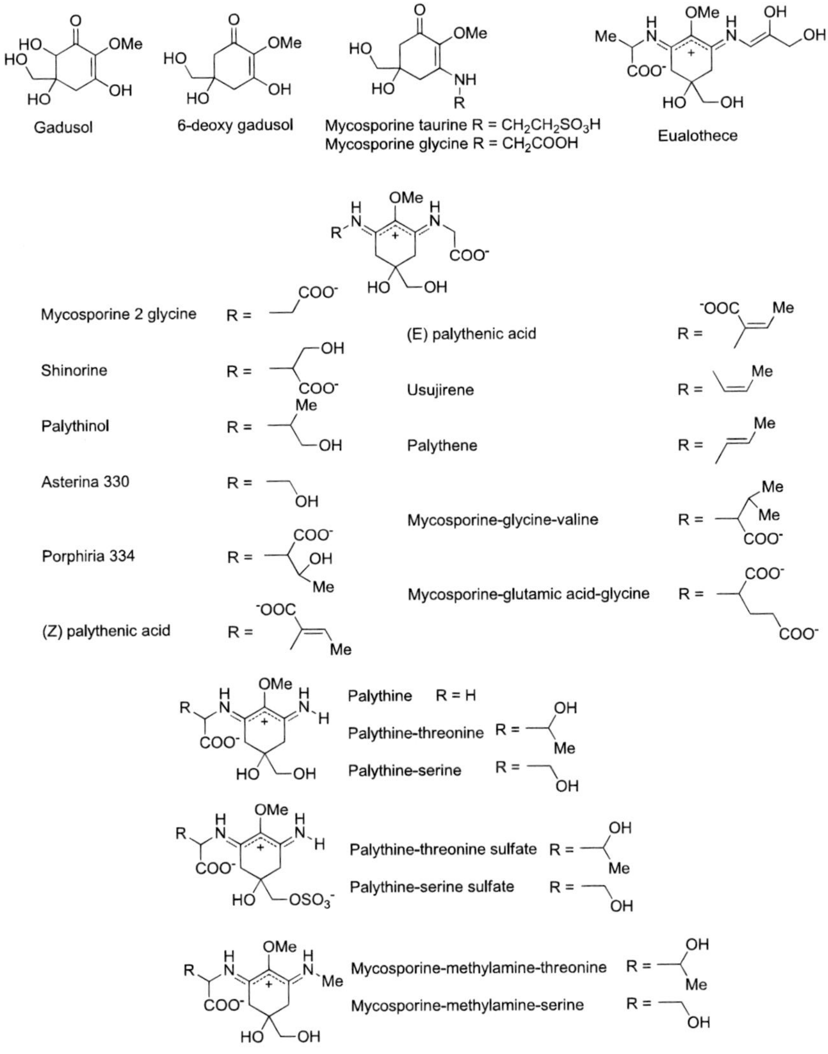

Figure 2. Gadusol, 6-deoxygadusol, oxo-mycosporines and imino-mycosporines in marine organisms.

In fungi, the only amino acids present in mycosporines are serine (including its α-amino alcohol, serinol) and the related pairs glutamine-glutaminol and glutamic acid- glutamicol,[11] with the exception of mycosporine alanine. The first isolated and elucidated structure was the mycosporine isolated from *BasidiomycetesSteriumHirutum*[33] and elucidated in 1976. The mycosporine isolated from *Botrytis cinera*[12] presents a derivative of glutamic acid and appears both as open chain and ring form (Figure 1).

In contrast with fungal mycosporines, in marine organisms and algae the predominant structure is the aminocyclohexenimine ring, with only two exceptions of aminocyclohexenone. These metabolites are imine derivates of mycosporines (enamine-imine) and are also called mycosporine-like amino acids (MAAs). These compounds present a high absorption band between 320-360 nm. The more common amino acid in these compounds is glycine, but a lot of variations are observed in the *ca.* 30 MAAs found until now (Figure 2).

In many cases, MAAs appear in nature as a combination with different amino acids. For instance, glycine is usually located at C3 and a second amino acid, an amino alcohol or an enaminone may be located in C1. However, in some coral species, glycine has been replaced by methyl amine (mycosporine-methyl-amine-serine, mycosporine-methylamine-threonine) or amine group (palythine, palythine-serine and palythine-threonine).[34] In addition, gadusol and 6-deoxygadusol have been found in some coral species.[35] Finally, some MAAs also contain sulfate esters,[36] such as palythine threonine sulfate and palythine serine sulfate. Recently, a new rare MAA was isolated[37] from the unicellular cyanobacterium*Euhalothece* and the Nori antioxidant compound was synthesized[38] by heat treatment of porphyra-334 (Figure 3).

The detailed structure of this type of compounds has been also obtained by diverse characterization techniques. There are some examples of X-ray diffraction studies on the total elucidation of porphyra-334 and palythine.[39]

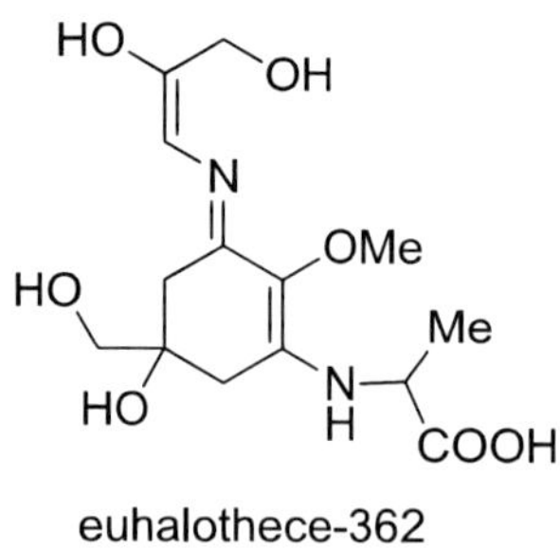

euhalothece-362

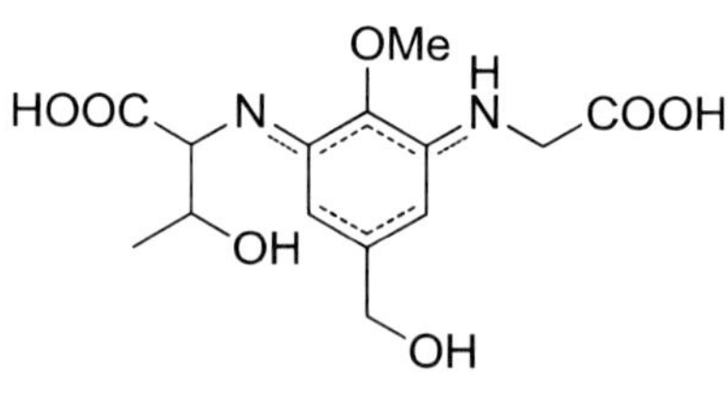

Nori antioxidant compound

Figure 3. New types of mycosporines recently isolated or synthesized.

5. Biological Roles

The presence of MAAs in different types of organisms, both marine and terrestrial, together with the widespread distribution along diverse ecosystems suggests a rather general and relevant biological role. Indeed, many different roles have been assigned to MAAs based in their unique features. For instance, it has been proposed that these compounds could help in the control of the intracellular osmotic pressure, as MAAs biosynthesis can be induced by osmotic shock.[40] However, it seems that the contribution of MAAs as compatible solutes is quantitatively small.[41] Also, their intervention in sexual gametes has been invoked as high proportions of these compounds have been found in embryos of shrimp.[19] According to this, MAA's concentration would be used as an indicator of ovarian maturation.[42] Another possible function is their role as accessorial pigments, precursor of photosynthetic pigments[43] or with a photosynthetic role. Indirect proof for this function is the fluorescent emission of the MAAs at 430-450 nm, which could excite chlorophyll and promote the use of UV light in the photosynthesis. However, recent studies conclude that MAA don't transfer efficiently energy to the chlorophyll A and, in conclusion, MAAs don't take part in photosynthesis.[44]

Antioxidant effects were found in mycosporine-glycine[45] in 1995. This compound is an efficient antioxidant that can protect biological systems by quenching oxygen. However, it seems that this property is not general for MAAs.[45] Antioxidant properties may be in relation with the quenching capacity of MAAs and their photoprotective role.[44] Recently, the activity of mycosporine-glutaminol-glucoside has also been studied.[46] This compound is both an efficient antioxidant and photoprotective metabolite as it can act as a singlet oxygen quencher.

Structurally related to MAAs (see Figure 2), gadusol appears in different locations of the physiology of diverse organisms. It seems that MAAs and gadusol share some biochemical transformations and functions. Especially, gadusol features a considerable antioxidant capacity, redox properties similar to ascorbic acid and the ability to quench chain reactions induced by peroxy radicals.[42] These relations between MAAs and gadusol have induced the suggestion that gadusol may be the evolutionary precursor of MAAs. The stronger antioxidant capacity of gadusol, together with the photoprotection in more energetic regions of the UV spectrum could be linked to the characteristics of the early atmosphere, with strong UVC radiation and an increased oxidative nature. Recent studies have shown[47] that the enolate form of gadusol may be responsible for the antioxidant capacities of this compound. At physiological pH = 7, the main form in aqueous solution of gadusol is the negatively charged enolate. This enolate form may undergo electron transfer reactions with different triplet sensitizers in water. In fact, the rose bengal triplet state is quenched by gadusolate through a collisional mechanism with a rate constant below the diffusion limit. This reductive quenching mechanism is considered to be responsible for the antioxidant capacity of gadusol *in vivo*. In living cells, the presence of critical compounds such as riboflavin and derivatives are related with DNA damage through photosensitized reactions. Gadusol is capable of efficiently contributing to diminish this potentially harmful reactions through quenching of the triplet of riboflavin.

6. Photoprotective Role

Beyond these suggested biological roles, the principal function of MAAs is the photoprotective role, acting as sunscreens helping microorganisms to cope with UV-radiation. Several photophysical and photochemical features of MAAs allowed to attribute a photoprotective role to these metabolites. MAAs show a strong absorption in the region of the spectrum potentially more dangerous to the organisms. Also, these compounds have shown a very high photostability both *in vivo* and *in vitro*. Furthermore, MAAs studied showed no relevant emission. Thus, the main deactivation pathway for MAAs is the thermal dissipation of the light energy without any relevant radiative or reactive deactivation channel, as it is expected for efficient sunscreens. Besides, direct information on the sunscreen capabilities of MAAs has been obtained in dinoflagellates.[48] In the case of organisms capable of synthesizing MAAs, the production of these metabolites is enhanced in the presence of intense solar radiation.[42] In the rest of the cases, MAAs can be obtained through the diet and are accumulated in the tissues sensitive to UV-radiation such as eye lenses.

6.1. Porphyra-334

Although the photoprotective role of MAAs was first suggested several decades ago and direct and indirect proof has been accumulating since then, the photophysical properties of specific compounds have been only lately studied in detail. Thus, from a photochemical point of view, the critical factors controlling the sunscreen capabilities, precise data on the deactivation processes and mechanistic aspects at the molecular level are only available for selected compounds. One of the better known MAAs is porphyra-334 (Figure 4). This

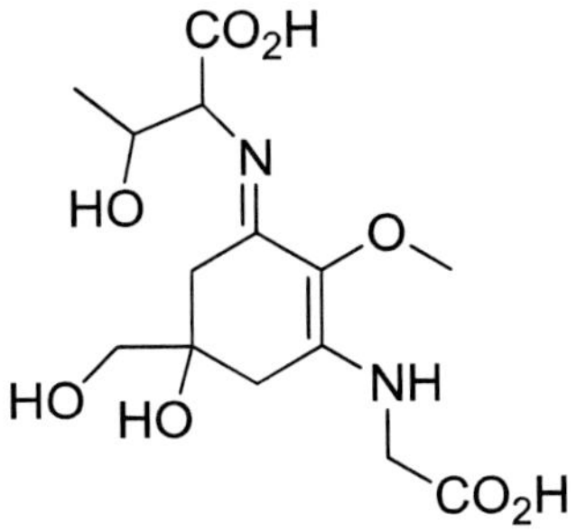

Figure 4. Molecular structure of porphira-334.

MAA was first isolated[30] from the red alga *Porphyratenera* and features a strong absorption band with a maximum located at 334 nm. However, it took more than 20 years to thoroughly explore the excited-state properties.[49] Porphyra-334 features an excited-state short lifetime of 0.4 ns and no transient species were located. Interestingly, this compound has very low quantum yields of fluorescence ($1.6*10^{-3}$), triplet formation ($<5*10^{-2}$) and photodecomposition ($2*10^{-4}$). Also, it is very stable in aqueous solution in the dark at 45°C. All this data are in good agreement with a very fast and efficient internal conversion process without the formation of any intermediate. Photoacousticcalorimetry measurements allowed to quantify the magnitude of non-radiative deactivation pathways of the excited species.[50] It was found that around 98% of the absorbed energy was dissipated as heat to the environment. Thus, this compound has all the features that should be desirable for an ideal sunscreen. In this case, the photophysical and photochemical properties clearly show that the strong absorption in a relevant region of the spectrum together with a very efficient energy dissipation as heat contribute to the protection of the cell. Also, as the formation of triplet species or radicals which could subsequently damage the

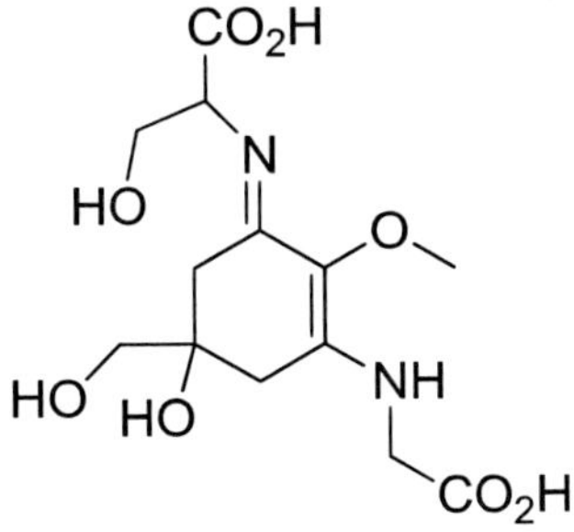

Figure 5. Molecular structure of shinorine.

organism is very limited, there is no need of indirect mechanisms to protect living tissues.

To check the sunscreen capabilities, the efficiency of a methanolic extract of porphyra-334 was tested against two commercial sun care products.[51] It was found that this compound can be used as a UV-A protecting sunscreen with a sun protection factor (SPF) five times more than the commercial products. Furthermore, as no reactive species are created after light absorption, no phototoxic effect on living organisms is expected.

6.2. Shinorine

Structurally very similar to porphira-334, the MAA shinorine is usually found in the same organisms but sometimes in bigger amounts (Figure 5). It has been obtained from adult sea urchins eggs as methanolic extracts.[52]

Shinorine was found to be photostable for at least 24 hours of continuous irradiation *in vivo*. Also, the photoprotective role of this MAA in the urchin eggs was also proven as it was found a relation between the intracellular MAA concentration and the UV-induced cleavage of embryos. Besides, neither photodegradation nor radical formation were detected after long-term exposure to UV-light *in vitro*. Detailed photophysicsmeasurements[50] of shinorine also showed that, as porphira-334, this metabolite efficiently dissipates light energy into heat. Shinorine has too a very low quantum yield of fluorescence ($1.6*10^{-4}$), a short fluorescence lifetime (0.35 ns), and small quantum yields for triplet formation ($<5*10^{-2}$) and photodecomposition ($3.1*10^{-4}$).

6.3. Palythine

Palythine is the MAA with the simplest structure (see Figure 6). It was first isolated[53] from the zoanthid*Palythoatuberculosa*. The structure of palythine and other related compounds (mycosporine-glycine, palythinol and palythine) obtained from the same source was later proposed.[22a] A very detailed photochemical analysis of palythine has been performed.[54] This MAA features a strong absorption at 320 nm (ε = 36200 M^{-1} cm^{-1}) and it is very stable in aqueous neutral solutions at 45°C. Also, palythine is very photostable as its photodecomposition quantum yield is $1.2*10^{-5}$. The lack of measurable luminescence implies a very efficient deactivation pathway. Specifically, photoacousticcalorimetry shows that around 90% of the light energy is transferred to the medium as heat.

Different derivatives of palythine have been isolated from natural sources and studied.[36] Sulfate esters of palythine were isolated from the reef-building coral *Stylophorapistillata*. These derivatives show the same absorption as palythine but are highly polar. Also, substitution of the palythine with different amino acids as threonine or serine has been also reported.[55]

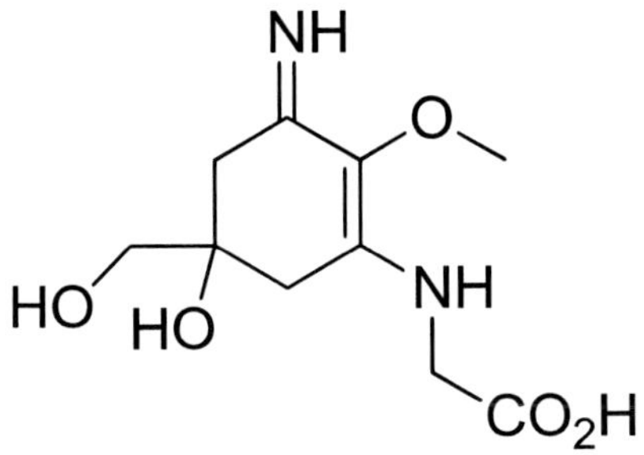

Figure 6. Molecular structure of palythine.

6.4. Deactivation Process at the Molecular Level

As explained above, the photoprotective role of these compounds has been invoked, together with many other roles such as the antioxidant activity, osmotic regulation, reproduction control and use as nitrogen reservoirs. Indirect evidence for the sunscreen capabilities comes from their physicochemical properties such as the strong UV-absorption, high photostability and lack of fluorescence of these compounds.[49] Although MAAs have been extensively studied from a biological point of view, little is known on the molecular basis of the UV protection. Some chemical properties of these compounds may be relevant for the photoprotective capabilities. MAAs are low-molecular-weight water-soluble molecules. Also, the UV spectra for these compounds feature only one band between 310-360 nm with strong absorption (ε = 30000-40000). In most cases, MAAs are obtained from natural sources and sometimes purified through long processes. This has contributed to some degree of ambiguity in the proposed chemical structures of these compounds. For instance, it is not clear whether the neutral or protonated forms of MAAs are active *in vivo*. Thus, the first step to understand the molecular aspects of the photoprotective capabilities is to clarify the chemical structure. A detailed computational study[56] was carried out using the CASPT2//CASSCF strategy.[57] Palythine was chosen as a model compound for MAA as it has the simplest structure (Figure 6). Two different structures could be relevant for the *in vivo* sunscreen activity, the protonated and the neutral forms (Figure 7).

To test the relevance of each structure, the computed UV spectrum for each form was checked against the available experimental data. Palythine features an absorption band at 320 nm with ε = 36200. The computed spectrum in gas phase for the neutral form showed a band at 256 nm, while the protonated form showed a band at 280 nm. Also, the protonated form features a much stronger absorption. Thus, the spectroscopic properties of the protonated form are in better agreement with the experimental data. As MAAs are water-soluble species, the UV spectra were also computed including the solvent effect. In both cases, absorptions and oscillator strengths remained very similar. Thus, it seems that the protonated imino species (see Figure 7, right) is responsible for the strong UV absorption found *in vivo*.

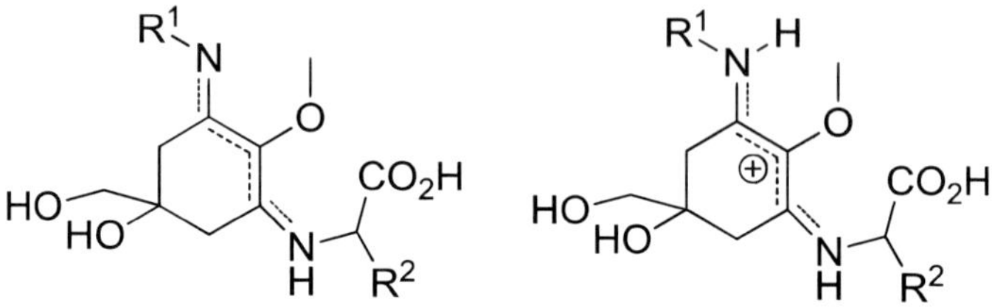

Figure 7. Neutral (left) and protonated forms of palythine.

In order to get a deeper insight into the molecular mechanism of photoprotection, the reactive paths after light absorption for both species were computed. In the case of the neutral form, two different absorption bands were computed corresponding with transitions to the excited states S_1 and S_2. The minimum energy paths (MEPs) for the reaction along both potential energy surfaces were computed. Analysis of the results showed that very fast deactivation may occur after population of any of the two excited states. Then, decay to the ground state can be easily achieved due to the presence of two low-energy conical intersection points (CI) accessible along the MEPs. These CI points connect the states S_2/S_1 and S_1/S_0 and they can be reached by quite small geometrical changes at low energies in the energy surface. The geometry of the more relevant S_1/S_0 CI point is shown in Figure 8.

It can be seen that the main geometrical deformation in this structure is the distortion of the chromophore. The N=C-C=C moiety responsible for the UV-absorption together with the substituentsareplanar in the ground state. However, the structure in the CI point shows how the groups located in the C=C double bond move away from planarity. This movement contributes to increase the energy of the ground state which crosses with the excited state at the CI point. Once reached the ground state, relaxation leads to the recovery of the ground state minimum, but no photoproducts were formed. This is in agreement with the high photostability experimentally found for palythine.

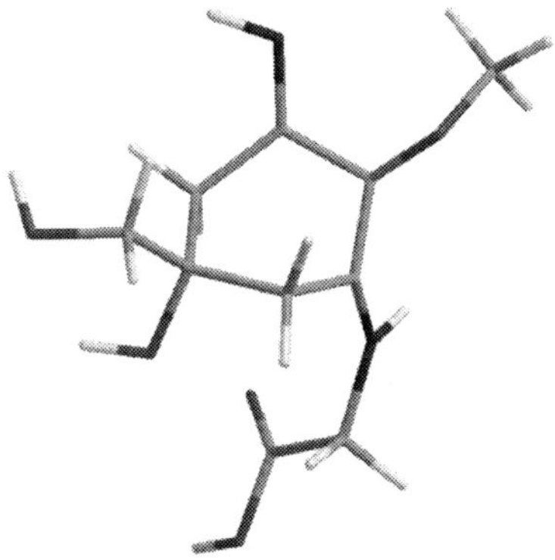

Figure 8. Geometry of the S_1/S_0 CI point for the neutral form of palythine.

Also, the protonated form of palythine was computed. After S_1 population, fast relaxation on the excited state energy surface was found as well. Analysis of the S_1/S_0 CI geometry (Figure 9) shows that the main geometrical modification corresponds with an out-of-plane movement of the imine moiety. This result implies that the deactivation pathway is more complex than a simple C=N bond isomerization, as previously suggested. Relaxation on the ground state recovers the palythine in its minimum together with a conformer and no photoproduct is formed after direct irradiation. Thus, this form is expected also to be very photostable.

From the reported theoretical data, it seems that the protonated form of palythine is responsible for the main absorption *in vivo*, although both neutral and protonated species share some photophysical and photochemical features that confer palythine its physiological activity. Thus, for the protective capabilities of palythine (and probably the rest of MAAs and related compounds) possibly the neutral and protonated forms are relevant. This fact seems reasonable as both species could be in equilibrium at physiological pH. Both forms feature a rapid deactivation which very effectively dissipates light energy into heat. This is especially important in the protonated form, due to the strong absorption band. Besides, the absence of any intermediates along the photoreaction path is in agreement with the lack of fluorescence in this compound. This fact, together with the recovery of the initial form after irradiation, offers a high photostability to this compound which in turn allows for efficient energy wastage through subsequent photocycles. Thus, this metabolite is stable enough to act as a sunscreen for long periods of time.

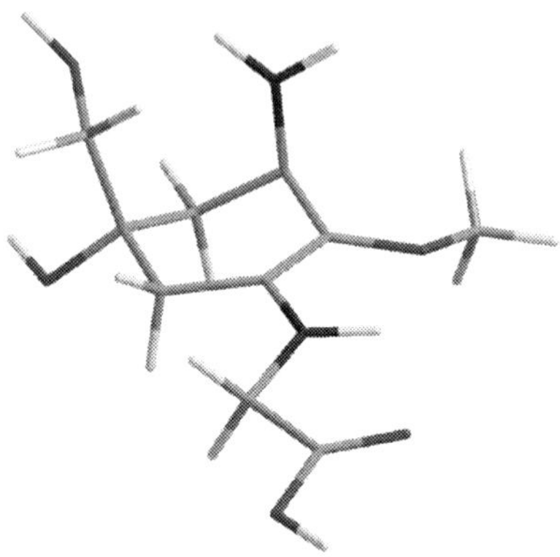

Figure 9. Geometry of the S_1/S_0 CI point for the protonated form of palythine.

The experimental and theoretical results gathered to date clearly prove that palythine has a very relevant photoprotective capability. Generalization to other MAAs seems reasonable as the modifications that appear in the general structure are only due to specific metabolic routes for each organism. These modifications only contribute to small variations in the photophysical properties, such as the maximum of the absorption band. Thus, at a molecular level, the potential of these compounds to protect living organisms from UV radiation seems evident. However, the role of MAAs as multipurpose secondary metabolites is still open.

7. Conclusions and Outlook

The role of MAAs in biological systems has been extensively studied since their discovery. However, many key features on the biosynthesis or the roles in different organisms have to be addressed yet. From the biological point of view, the location, evolutionary aspects or the biogenesis of MAAs in diverse organisms will require a detailed study before we could entirely understand the relevance of these metabolites and their functioning in living organisms. As these compounds have shown an impressive potential as antioxidants and photoprotective species, a comprehensive investigation of the chemical reasons behind this behavior will definitively shed some light on the mechanisms of action at the molecular level. Even more, MAAs and related compounds may be useful sources of inspiration for the development of new compounds with interesting properties for practical, commercial applications. Some steps in this direction have been already taken, especially in the field of MAA-based commercial sunscreens. The extraordinary tolerance of certain organisms to potentially damaging UV radiation has been related to the unique features of MAAs as sunscreens. Thus, the high absorption, high photostability and lack of reactive intermediates in these compounds have suggested their use as components of skin-care products. Several synthetic analogues of MAAS have been already synthesized and tested in diverse applications. Unfortunately, the derivatives tested to date do not conform to the requirements of the commercial products, mainly due to their high instability towards hydrolysis and oxidation. However, the extraordinary features of MAA and the wide biological distribution suggest that these compounds have the potential to encourage further efforts to develop relevant applications.

Acknowledgments

We thank the Spanish MINECO/FEDER (CTQ2011-24800), Comunidad Autónoma de La Rioja and Universidad de La Rioja for financial support. R. L. thanks the Spanish MECD for his fellowship.

References

[1] Bruce, V., Green, P. R. & Georgeson, M. A. (2003).*Visual Perception: Physiology, Psychology and Ecology*, 4th ed., Psychology Press, Hove.
[2] Hall, K. K. R. (1999).*Photosynthesis (Studies in Biology),* Vol. 6, Cambridge University Press, Cambridge.
[3] Häder, D.-P. & Lebert, M. (2001).*Photomovement*, Elsevier, Amsterdam.
[4] Wang, Y. & Casadevall, A. (1994).*Appl. Environ. Microbiol.*, 3864-3866.
[5] Rastogi, R. P., Kumar, A. R., Tyagi, M. B. & Sinha, R. P. (2010).*J. Nucleic Acids*, 592980.
[6] Young, A. J. (1991).*Physiol. Plant.*, 83, 702-708.
[7] Garcia-Pichel, F. & Castenholz, R. W. (1991).*J. Phycol.*, 27, 395-409.
[8] Butler, M. J. & Day, A. W. (1998).*Can. J. Microbiol.*, 44, 1115-1136.
[9] Britton, G., Liaaen-Jensen, S. & Pfander, H. (2004).*Carotenoids Handbook*, Birkhäuser Verlag, Basel.
[10] Nakamura, H., Kobayashi, J. & Hirata, Y. (1982).*J. Chromatogr.*, 250, 113.

[11] Arpin, N. & Bouillant, M. L. (1981). in *The Fungal Spore: Morphogenetic Controls* (Eds.: G. Turian, H. R. Hohl), Academic Press, London, 435-454.

[12] Arpin, N., Favre-Bonvin, J. & Thivend, S. (1977). *Tetrahedron Lett.*, 10, 819.

[13] Karentz, D., McEuen, F. S., Land, M. C. &Dunlap, W. C. (1991).*Mar. Biol.*, 108, 157.

[14] Weste, G. (1970). *Aust. J. Bot.*, 18, 11.

[15] Brook, P. J. (1981).*N. Z. J. Bot*, 19, 299.

[16] Dionisio-Sese, M. L., Ishikura, M., Maruyama, T. & Miyachi, S. (1997).*Mar. Biol.*, 128, 455.

[17] Arai, T., Nishijima, M., Adachi, K. & Sano, H. (1992).*MBI Report*, Marine Biotechnology Institute, Tokio.

[18] Dunlap, W. D. & Chalker, B. E. (1986).*Coral Reefs*, 5, 155.

[19] Grant, P. T., Middleton, C., Plack, P. A. & Thompson., R. H. (1985).*Comp. Biochem. Physiol.*, 80B, 755-759.

[20] Dunlap, W. C., Williams, D. M., Chalker, B. E. & Banaszak, A. T. (1989).*Comp. Biochem. Physiol.*, 93B, 601.

[21] Drollet, J. H., Glaziou, P. & Martin, P. M. V. (1993).*Mar. Biol.*, 115, 263.

[22] a) Hirata, Y., Uemura, D., Ueda, K. & Takano, S. (1979).*Pure Appl. Chem.*, *51*, 1875. b) Favre-Bonvin, J., Bernillon, J., Salin, N. & Arpin, N. (1990).*Phytochemistry*, 29, 81.

[23] Haslam, E. (1993). in *Shikimic Acid: Metabolism and Metabolites*, Wiley, Chichester.

[24] Bentley, R. (1990).*Crit. Rev. Biochem. Mol. Biol.*, 25, 307-384.

[25] a) Balskus, E. P. & Walsh., C. T. (2010).*Science*, 329, 1653. b) Rosic, N. N. & Dove, S. (2011).*App. Env. Microbiol.*, 77, 8478-8486.

[26] Portwich, A. & Garcia-Pichel, F. (2003).*Phycologia*, 42, 384-392.

[27] Misonou, T., Saitoh, J., Oshiba, S., Tokitomo, Y., Maegawa, M., Inoue, Y., Hori, H. & Sakurai, T. (2003).*Mar. Biol.*, 5, 194-200.

[28] Carreto, J. I., Carignan, M. O. & Montoya, N. G. (2005).*Mar. Biol.*, 146, 237.

[29] García, P. E., Diéguez, M. C., Ferraro, M. A., Zagarese, H. E. & Pérez, A. P. (2010).*Photochem. Photobiol.*, 86, 353-359.

[30] Takano, S., Nakanishi, A., Uemura, D. & Hirata, Y. (1979).*Chem. Lett.*, 419.

[31] Gleeson, D. F. &Wellington, G. M. (1993).*Nature*, 365, 286.

[32] Bernillon, J., Bouillant, M. L., Pittet, J. L., Favre-Bonvin, J. & Arpin, N. (1984).*Phytochemistry*, 23, 1083.

[33] Favre-Bonvin, J., Arpin, N. & Brevard, C. (1976).*Can. J. Chem.*, 54, 1105.

[34] Teai, T., Raharivelomanana, P., Bianchini, J. P., Faura, R., Martín, P. M. V. & Cambon, A. (1997).*Tetrahedron Lett.*, 38, 5799-5800.
[35] Arbeloa, E. M., Carignan, M. O., Acuña, F. H., Churio, M. S. & Carreto, J. I. (2010).*Comp.Biochem. Physiol.* B, 156, 216-221.
[36] Wu, J. J., Chalker, B. E. &Rideout, J. A. (1997).*Tetrahedron Lett.*, 38, 2525.
[37] Volkmann, M., Gorbushina, A. A., Kedar, L. & Oren., A. (2006). *FEMS Microbiol. Lett.*, 258, 50.
[38] Yoshiki, M., Tsuge, K., Tsuruta, Y., Yoshimura, T., Koganemaru, K., Sumi, T., Matsui, T. & Matsumoto, K. (2009). *Food Chem.*, 113, 1127.
[39] Klisch, M., Richter, P., Häder, D.-P. & Bauer, W. (2007).*Helv. Chim. Acta*, 90, 488-511.
[40] Kogel, T., Gostincar, C., Volkmann, M., Gorbushina, A. A. & Gunde-Cimerman, N. (2006).*Environ. Chem.*, 3, 105-110.
[41] Portwich, A. & García-Pichel, F. (1999).*Arch. Microbiol.*, 172, 187-192.
[42] Shick, J. M. & Dunlap, W. C. (2002).*Ann. Rev. Phys.*, 64, 223-262.
[43] Price, J. H. & Forrest, H. S. (1993).*Comp. Biochem. Physiol.*, 30, 879-888.
[44] Moisant, T. A. & Mitchell, B. G. (2001).*Mar. Biol.*, 138, 217-227.
[45] Dunlap, W. C. & Yamamoto, Y. (1995). *Comp. Biochem. Physiol. B*, 112, 105.
[46] Moliné, M., Arbeloa, E. M., Flores, M. R., Libkind, D., Farías, M. E., Bertolotti, S. G., Churio, M. S. & Broock, M. R. (2011).*Rad. Res.*, 175, 44-50.
[47] Arbeloa, E. M., Bertolotti, S. G. & Churio, M. S. (2011).*Photochem. Photobiol. Sci.*, 10, 133.
[48] Neale, P. J., Banaszak, A. T. & Jarriel, C. R. (1998)*J. Phycol.*, 34, 928-938.
[49] Conde, F. R., Churio, M. S. &Previtali, C. M. (2000).*J. Photochem. Photobiol. B*, 56, 139-144.
[50] Conde, F. R., Churio, M. S. & Previtali, C. M. (2004).*Photochem. Photobiol. Sci.*, 3, 960-967.
[51] Torres, A., Enk, C. D., Hochberg, M. & Srebnik, M. (2006). *Photochem. Photobiol. Sci.*, 5.
[52] Adams, N. L. & Shick, J. M. (1996).*Photochem. Photobiol.*, 64, 149-158.
[53] Takano, S., Uemura, D. & Hirata, Y. (1978).*Tetrahedron Lett.*, 26, 2299-2300.
[54] Conde, F. R., Churio, M. S. & Previtali, C. M. (2007).*Photochem. Photobiol. Sci.*, 6, 669-674.
[55] Carignan, M. O., Cardozo, K. H., Oliveira-Silva, D., Colepicolo, P., Carreto, J. I. (2009).*J. Photochem. Photobiol. B*, 94, 191-200.

[56] Sampedro, D. (2001).*Phys. Chem. Chem. Phys.*, 13, 5584-5586.
[57] Sampedro, D. (2011). in *Photochemistry: UV/VIS Spectroscopy, Photochemical Reactions and Photosynthesis* (Eds.: K. J. Maes, J. M. Willems), Nova Science Publishers.

In: Photochemistry: New Research
Editors: A. B. Smith et al, pp. 21-35
ISBN: 978-1-62808-615-7

Chapter 2

BICARBONATE-ENHANCED PHOTOCATALYTIC DEGRADATION OF PHENOL:COMPETITION WITH EXCESS ORGANIC MATTER

Radharani Das[1,2], Claudio Minero[2], Valter Maurino[2] and Davide Vione[2,*]

[1]Department of Chemical Engineering,
Haldia Institute of Technology, ICARE Complex, Haldia, India
[2]Dipartimento di Chimica Analitica, Università di Torino, Torino, Italy

Abstract

Sodium bicarbonate is able to inhibit the TiO2-assisted photocatalytic degradation of phenol in aqueous solution. However, in the presence of excess 2-propanol the transformation of phenol can be enhanced by bicarbonate. A likely explanation is that TiO2 under irradiation oxidises the carbonate and bicarbonate anions to the carbonate radical (CO3−•), which reacts selectively with phenol and does not react (or reacts only to a very limited extent) with 2-propanol. NaHCO3 would, therefore, shift reactive species from the transformation of 2-propanol to that of phenol, which could increase the degradation rate of phenol in the presence of bicarbonate. The enhancement effect by bicarbonate would be operational in the presence of organic matter in excess, and at bicarbonate levels that would be typical of waters with elevated alkalinity.

* E-mail address: davide.vione@unito.it., Tel: +39-011-6705296, Fax: +39-011-6707615, http://naturali.campusnet.unito.it/cgi.bin/docenti.pl/Show?_id= vione (Corresponding Author)

Keywords: Inorganic carbon; dissolved organic matter; photocatalysis; titanium dioxide; water decontamination

Introduction

Heterogeneous photocatalysis is one of the most promising advanced oxidation processes for water and wastewater decontamination [1-4]. It is based on the irradiation of semiconductor oxides, which induces the promotion of electrons from the valence to the conduction band of the oxide, leaving electron vacancies (holes) in the valence band. The valence-band holes (h^+_{VB}) can migrate toward the photocatalyst surface, where they are entrapped to form oxidised species on the surface or the immediate sub-surface layer of the oxide. These species induce the oxidative transformation of dissolved molecules, including organic pollutants. Conduction-band electrons (e^-_{CB}) entrapped by surface groups form reducing species, which can be scavenged by dissolved oxygen to give the superoxide radical anion. This is the first step of the reduction process of O_2 to H_2O via H_2O_2. Interestingly, the one-electron reduction of H_2O_2 or its UV photolysis yield the hydroxyl radical $^\bullet OH$, which is a very reactive and powerful oxidant and can also take part to the degradation of organic contaminants [1]. In some cases (including the Fe(III) and Mn(III,IV) (hydr)oxides) the surface entrapment of e^-_{CB} can yield soluble species, such as Fe^{2+} and Mn^{2+}. As a consequence, photocorrosion of the semiconductor oxide can take place. Such a phenomenon is not desired for the applications of heterogeneous photocatalysis, for which photostable materials are generally required [5].

Titanium dioxide (TiO_2) is by far the most frequently used material in photocatalysis. Reasons for its wide use include water insolubility, photostability, low cost, lack of toxicity, and obviously the elevated activity toward photodegradation processes, which is a distinguished feature of the anatase form of TiO_2. An important issue is that TiO_2 does not undergo detectable photocorrosion processes. In the case of anatase TiO_2, of which the well-known Degussa P25 is the most widely studied variety, the reactions that are induced by absorption of UV radiation are the following [1]:

$$TiO_2 + h\nu \rightarrow h^+_{VB} + e^-_{CB} \quad (1)$$

$$h^+_{VB} + (Ti^{IV}{-}OH^-)_{surf} \rightarrow (Ti^{IV}{-}^\bullet OH)_{surf} \quad (2)$$

$$h^+_{VB} + (Ti^{IV}{-}O^{2-}{-}Ti^{IV})_{sub\text{-}surf} \rightarrow (Ti^{IV}{-}O^{-\bullet}{-}Ti^{IV})_{sub\text{-}surf} \quad (3)$$

$$e^-_{CB} + (Ti^{IV}{-}OH^-)_{surf} \rightarrow (Ti^{III}{-}OH^-)_{surf} \quad (4)$$

Trapped electrons can react with dissolved O_2, while the holes trapped by surface and sub-surface groups are able to oxidise various compounds in solution via hydroxylation (operated by $(Ti^{IV}{-}^{\bullet}OH)_{surf}$) or charge transfer (by $(Ti^{IV}{-}O^{-\bullet}{-}Ti^{IV})_{sub\text{-}surf}$). Oxidative transformation processes can also be induced by the $^{\bullet}OH$ radicals that are formed in the pathway of O_2 reduction to H_2O. In competition with these reactions, recombination processes between trapped electrons and holes can occur with dissipation of thermal energy. Although recombination between surface- and sub-surface trapped electrons and holes is much slower than electron-hole recombination in the semiconductor bulk, these processes decrease the efficiency of photodegradation. Recombination reactions are usually prevailing in the absence of oxidisable species in solution. A decrease of the efficiency of photocatalytic reactions can also be induced by the reduction of partially oxidised organic radicals or intermediates, or by the oxidation of partially reduced species. In all these cases null cycles are operational, in which the organic substrates catalyse the recombination of electrons and holes with dissipation of heat. The recombination catalysed by organic compounds (S) is usually prevailing at elevated concentrations of the latter [6-8].

$$(Ti^{IV}{-}O^{-\bullet}{-}Ti^{IV})_{sub\text{-}surf} + S \rightarrow (Ti^{IV}{-}O^{2-}{-}Ti^{IV})_{sub\text{-}surf} + S^{+\bullet} \quad (5)$$

$$S^{+\bullet} + (Ti^{III}{-}OH^-)_{surf} \rightarrow S + (Ti^{IV}{-}OH^-)_{surf} \quad (6)$$

The majority of the studies into the photocatalytic degradation of organic compounds deals with the transformation of a single species in solution, to ensure the understanding of the reaction pathways. However, in practical applications of photocatalysis the scenario of a wastewater containing a single pollutant to be degraded is rather rare. Much more commonly one may expect the presence of inorganic anions and of several dissolved organic compounds, which might not contribute to pollution but can interfere with the degradation of the target substrate. While the separate effects of the presence of either inorganic anions or organic interfering agents on photocatalytic degradation processes have received wide attention [9], much less is known on the simultaneous presence of organic matter and inorganic anions. This is the focus of the present paper.

Bicarbonate is the most important anionic species in natural waters. Its effect on photocatalytic degradation processes has been the subject of various studies, and in the vast majority of cases it has been reported that HCO_3^- decreases the rate of photocatalytic reactions [10-17]. Like many inorganic anions, bicarbonate can compete with oxidisable substrates for the adsorption on the surface of TiO_2, with a negative effect on the degradation kinetics. Moreover, bicarbonate can be oxidised by irradiated TiO_2 to the radical $CO_3^{-\bullet}$, which is reactive toward organic and inorganic substrates [18] but much less than the species that are consumed for its production ($(Ti^{IV}-{}^{\bullet}OH)_{surf}$ and $(Ti^{IV}-O^{-\bullet}-Ti^{IV})_{sub\text{-}surf}$) [19]. Interestingly, in a minority of cases it has been observed that bicarbonate enhances the photocatalytic degradation processes [20]. It is very likely that such an enhancement is due to the basification of the solution, which can increase the surface density of the $(Ti^{IV}-OH^{-})_{surf}$ groups and, therefore, of $(Ti^{IV}-{}^{\bullet}OH)_{surf}$. Basification also increases the concentration of the conjugated bases of the substrates that undergo acid-baseequilibria, and these basic species are generally easier to oxidise than the corresponding acidic ones.

The presence of organic compounds is usually able to inhibit the photocatalytic degradation processes [9,20], because of the scavenging of reactive radical species. Considering that the oxidative degradation can take place either by hydroxylation (involving $(Ti^{IV}-{}^{\bullet}OH)_{surf}$) or by electron transfer (by $(Ti^{IV}-O^{-\bullet}-Ti^{IV})_{sub\text{-}surf}$), and that for some compounds one of the two pathways is strongly prevalent, the mutual interference might be very limited if the substrate and the potential interfering agent follow different transformation mechanisms [21,22]. However, in many cases the interfering species would be natural dissolved organic matter, and the wide variety of its constituents would ensure that all the degradation pathways undergo inhibition [20].

The simultaneous presence of inhibitors such as organic matter and bicarbonate can be expected to decrease the rates of the photocatalytic reactions, but the two effects might not be additive. For instance, in the photochemistry of natural waters the carbonate radical can play a significant role despite its lower reactivity compared to ${}^{\bullet}OH$. Indeed, $CO_3^{-\bullet}$ is more selective than ${}^{\bullet}OH$ for the reaction with certain compounds, and it is scavenged by dissolved organic matter to a far lesser extent [18,23,24]. To our knowledge, no study has been carried out so far into the effect of bicarbonate on the photodegradation by TiO_2 of organic substrates, in the presence of organic compounds that interfere with the process. The present work studied the photocatalytic degradation of phenol in the presence of 2-propanol, which is an inhibitor of the photocatalytic reactions [25]. Phenol is a widespread contaminant that can be found in wastewater from several processes, including petroleum refining, coal gasification and liquefaction, and the production of steel, dyes, synthetic resins and pesticides [26-28]. We show, for the first time to our knowledge, that under inhibition of photocatalytic processes by interfering organic material, bicarbonate may enhance the degradation reactions of phenol. In contrast, bicarbonate inhibits the photocatalytic degradation of phenol in the absence of organic matter.

Experimental Section

Reagents and Materials

Phenol (purity grade 99%), catechol (99%), hydroquinone (99%), resorcinol (99%), 1,4-benzoquinone (98%), 4-phenoxyphenol (99%), $NaHCO_3$ (>99%) and H_3PO_4 (85%) were purchased from Aldrich, 2-propanol (LiChrosolv gradient grade) from VWR Int., acetonitrile (Supergradient grade) from Scharlau. TiO_2 P25 was a gift from Degussa.

Irradiation Experiments

Irradiation of the TiO_2 suspensions (5 mL total volume) was carried out in Pyrex glass cells (4.0 cm diameter, 2.3 cm height), under a 100 W Philips TL01 UV-Vis lamp. The suspensions were illuminated by a UV photon flux of 3.3×10^{-6} Einstein L^{-1} s^{-1}, corresponding to a UV irradiance of 5.6 W m^{-2}, measured with a CO.FO.ME.GRA. (Milan, Italy) power meter (290-400 nm). The cells were magnetically stirred during irradiation. Figure 1 reports the emission spectrum of the lamp, measured with an Ocean Optics SD2000 CCD spectrophotometer, and the attenuance spectrum (absorption + scattering) of Degussa P25 TiO_2, taken with a Varian Cary 100 Scan UV-Vis spectrophotometer. The TiO_2 stock suspension (0.5 g L^{-1}) was sonicated before irradiation. After the scheduled irradiation time the suspensions were filtered on syringe-fitted Millipore Millex HA filter membranes (cellulose acetate, 0.45 μm pore diameter), and analysed by High-Performance Liquid Chromatography coupled with Diode Array Detection (HPLC-DAD).

The adopted instrument was a VWR-Hitachi HPLC equipped with L-2200 autosampler, L-2130 quaternary pump for low-pressure gradients, L-2300 column oven, and L-2455 DAD detector. The column used was a LiChroCART RP18 (VWR Int., 125×4 mm), packed with LiChrospher 100 RP 18 material (5 μm diameter). The isocratic eluent was a mixture of acetonitrile and aqueous H_3PO_4. The DAD detector was set to continuously acquire signal from 200 to 300 nm.

The elution conditions and the quantification wavelengths for phenol and its monitored transformation intermediates were as follows (%CH_3CN, % aqueous H_3PO_4, retention time in minutes, wavelength in nm): phenol (20, 80, 4.9, 210), catechol (20, 80, 2.7, 276), hydroquinone (20, 80, 1.7, 225), resorcinol (20, 80, 2.1, 210), 1,4-benzoquinone (20, 80, 2.6, 245), 4-phenoxyphenol (40, 60, 6.6, 210).

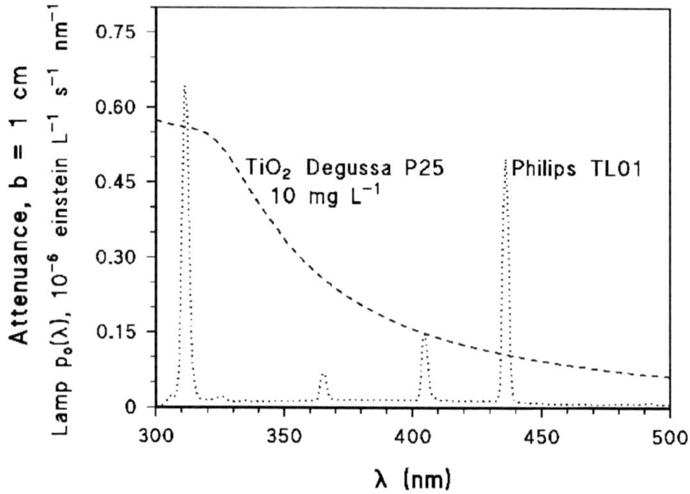

Figure 1. Emission spectrum ($p_o(\lambda)$) of the adopted UV-Vis lamp (Philips TL01, 100 W). Attenuance spectrum of TiO_2 Degussa P25 (10 mg L^{-1} loading, optical path length b = 1 cm).

Figure 2 reports as an example the DAD signal for phenol and its intermediates after 90 min irradiation of 0.1 mM initial phenol in the presence of 0.20 g L^{-1} TiO_2, 0.01 M 2-propanol, and 0.03 M $NaHCO_3$.

The time evolution curves of phenol were fitted with pseudo-first order degradation equations of the kind $C_t = C_0 exp\ (-k \cdot t)$, where C_t is the molar concentration of the substrate at the time t, C_0 its initial concentration, and k the pseudo-first order degradation rate constant. The initial degradation rates were calculated as $k \cdot C_0$, with k expressed in [s^{-1}] and C_0 in [mol L^{-1}]. The error bounds to the rate data ($\mu \pm \sigma$) represent the goodness of the fit of the pseudo-first order curve to the experimental data (intra-series variability).

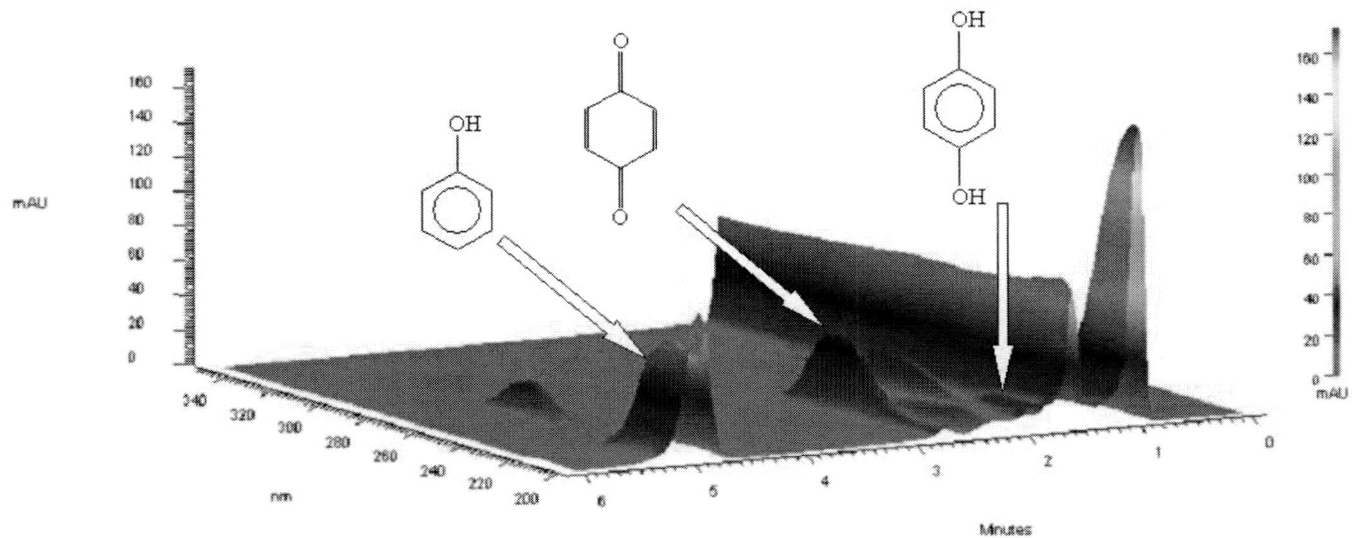

Figure 2. DAD chromatogram of phenol and its intermediates after 90 min irradiation of 0.1 mM initial phenol, in the presence of 0.20 g L^{-1} TiO_2 (Degussa P25), 0.01 M 2-propanol, and 0.03 M $NaHCO_3$.

Results and Discussion

Effect of Bicarbonate on the Degradation of Phenol with 2-Propanol

Figure 3 reports the initial degradation rate of 0.1 mM phenol as a function of the concentration of added $NaHCO_3$ (C_{NaHCO3}), upon UV-Vis irradiation of 0.20 g L^{-1} TiO_2, in the absence and in the presence of 2-propanol.

Focusing first on the system without propanol, it can be observed that bicarbonate substantially inhibits the photodegradation of phenol by TiO_2. The initial degradation rate of phenol is, in fact, about 35 times lower with 0.10 M $NaHCO_3$ than in the absence of bicarbonate. Two different phenomena mightaccount for the observed effect. The first is competition between phenol and bicarbonate for the adsorption on the surface of TiO_2, the second is the competition between phenol, HCO_3^- and CO_3^{2-} for the reaction with holes trapped by the surface or sub-surface groups of TiO_2 ($(Ti^{IV}-{}^{\bullet}OH)_{surf}$, $(Ti^{IV}-O^{-\bullet}-Ti^{IV})_{sub-surf}$). Phenol is poorly adsorbed on the surface of titania [29] and adsorption is, therefore, not a significant driving force for phenol photocatalytic degradation. As

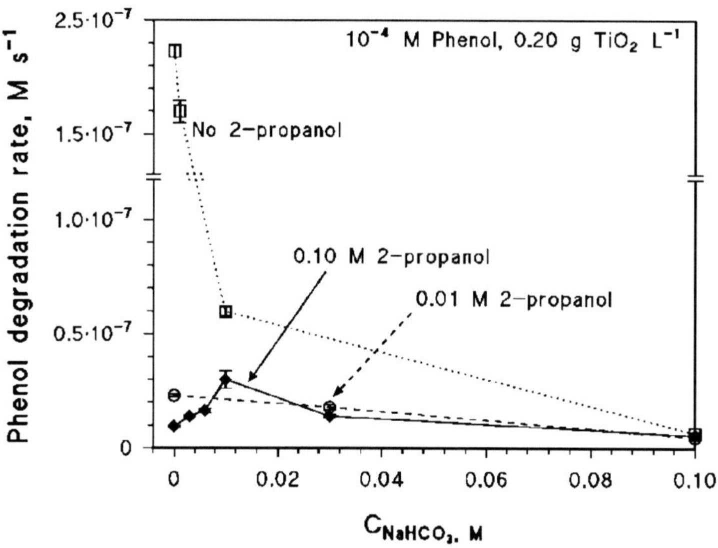

Figure 3. Initial degradation rates of 0.1 mM phenol upon UV-Vis irradiation of 0.20 g L^{-1} TiO_2 (Degussa P25), as a function of C_{NaHCO3}, in the presence of different concentrations of 2-propanol. Note the break in the Y-axis.

a consequence, the inhibition of photocatalytic degradation by bicarbonate is most likely accounted for by scavenging of reactive species with production of the less reactive radical $CO_3^{-\bullet}$. The carbonate radical can actually react with phenol, but the reaction rate constant is relatively low (around 10^7 M^{-1} s^{-1}) [19] and it could not compensate for the scavenging of the TiO_2oxidisingspeciesby carbonate and bicarbonate.

In the presence of 0.01 M 2-propanol the degradation of phenol is significantly inhibited compared to the system where the alcohol is absent (see Figure 3), because 2-propanol effectively competes with phenol for the oxidising species on the surface of TiO_2 [30]. Also in this case the addition of bicarbonate inhibits phenol transformation, but inhibition by bicarbonate is less marked than in the absence of 2-propanol. Indeed, the degradation rate of phenolwith 0.01 M 2-propanol is decreased by a factor of 5 when passing from no bicarbonate to 0.10 M $NaHCO_3$, compared to 35 times in the absence of 2-propanol.

The situation is completely different in the presence of 0.10 M 2-propanol: under such conditions, bicarbonate up to 0.01 M initial concentration enhances the photocatalytic degradation of phenol (Figure 3). From the rate data and the corresponding error bounds, one gets that phenol degradation is 3.2±0.6 ($\mu\pm\sigma$) times fasterwith 0.01 M $NaHCO_3$ than in the absence of bicarbonate. Figure 4 shows as an example the time evolution of 0.1 mM phenol in the presence of 0.10 M 2-propanol, without bicarbonate and with 0.01 M $NaHCO_3$ that enhances phenol degradation.

To account for these data, consider that the degradation rate of phenol is about 3.7 times lower in the presence of 0.1 mM phenol + 0.01 M $NaHCO_3$ than in the presence of 0.1 mM phenol alone. Moreover, the rate is 25 times lower with 0.1 mM phenol + 0.10 M 2-propanol than with 0.1 mM phenol (Figure 3). Under the hypothesis that the decrease of phenol degradation rate is mainly due to the scavenging of reactive species, one could hypothesise that 0.01 M $NaHCO_3$ is 2.7 timesmore efficient as scavenger than 0.1 mM phenol, and that 0.1 M 2-propanol is even 24 times more efficient than 0.1 mM phenol. In the system containing 0.1 mM phenol + 0.01 M $NaHCO_3$ + 0.1 M 2-propanol, if the above ratios are maintained, somewhat less than 4% of the reactive species would be scavenged by phenol, around 10% by $NaHCO_3$ (*i.e.*, by both HCO_3^- and CO_3^{2-}), and about 86% by 2-propanol. In the case of HCO_3^- and CO_3^{2-} the scavenging reaction would yield $CO_3^{-\bullet}$, which is reactive with phenol but shows negligible reactivity with 2-propanol [19]. The main reason for the differential reactivity is that the carbonate radical (differently from $^\bullet OH$ and $(Ti^{IV}-{}^\bullet OH)_{surf}$)) preferentially reacts *via* one-electron oxidation, toward which aliphatic compounds such as 2-propanol are virtually inactive. On the other hand, the delocalisedπ electrons of aromatic compounds such as phenol are more weakly bound to the molecule and can be abstracted more easily by monoelectronic oxidants [19,31].

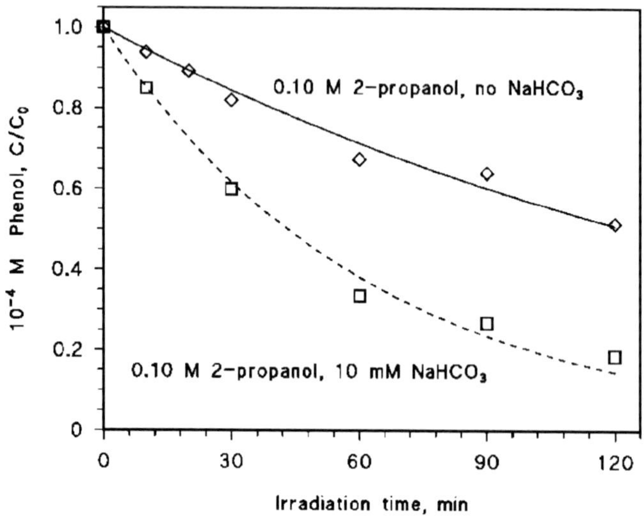

Figure 4. Time evolution of 0.1 mM phenol upon UV-Vis irradiation of 0.20 g L^{-1} TiO_2 (Degussa P25) in the presence of 0.10 M 2-propanol, without bicarbonate and with 10 mM $NaHCO_3$.

Because of the selective reactivity between phenol and $CO_3^{-\bullet}$, up to 14% of the reactive species produced by irradiation of TiO_2 could take part in the transformation of phenol in the presence of $NaHCO_3$ (that is, the ~4% that is scavenged by phenol plus the ~10% that is consumed by bicarbonate and carbonate to produce $CO_3^{-\bullet}$, which reacts almost exclusively with phenol). With 0.10 M 2-propanol but without bicarbonate, only ~4% of the photoproduced reactive species would react with 0.1 mM phenol. Therefore, the addition of bicarbonate could increase the degradation rate of phenol by up to 3.5 times, which is in agreement with the experimentally determined value of 3.2±0.6. In other words, the enhancement of phenol degradation by $NaHCO_3$ in the presence of 2-propanol would be caused by a shift of reactive species from the degradation of 2-propanol to that of phenol, carried out by HCO_3^- and CO_3^{2-} *via* $CO_3^{-\bullet}$.

The increasing degradation rate of phenol with increasing $NaHCO_3$ also suggests that dissolved inorganic carbon might enhance rather than inhibit the photocatalytic degradation of aromatic pollutants, if they are sufficiently reactive toward $CO_3^{-\bullet}$, in the presence of dissolved organic compounds that compete for photocatalytic degradation. Organic matter and bicarbonate, which could both inhibit degradation when present separately, would lead to faster photocatalytic transformation than could be foreseen by the sum of the separate effects.

Figure 3 also shows that the degradation rate of phenol with 0.10 M 2-propanol decreases above 0.01 M $NaHCO_3$. The decrease might be accounted for by the reduction of $CO_3^{-\bullet}$ to HCO_3^-/CO_3^{2-} by surface-bound electrons $((Ti^{III}-OH^-)_{surf})$. The reduction kinetics would be enhanced at elevated bicarbonate, and it would inhibit the rate of the photocatalytic reactions upon consumption of reactive species ($CO_3^{-\bullet}$ in this case). Such a process, involving the so-called back reactions, is well know to cause a decrease of the photocatalytic degradation rates of organic substrates at elevated concentration [6-8].

Among the transformation intermediates of phenol, a small amount of catechol was detected along with hydroquinone and 1,4-benzoquinone. The relatively low concentration of catechol in the aqueous phase is most likely accounted for by its higher adsorption on the surface of TiO_2, compared to hydroquinone and benzoquinone [30]. Figure 5 shows the time evolution of 1,4-benzoquinone upon irradiation of phenol (0.1 mM) + 2-propanol (0.1 M) + TiO_2 (0.20 g L^{-1}), for different concentration values of $NaHCO_3$. Under the reported conditions, bicarbonate enhances the degradation of phenol (see Figure 3).

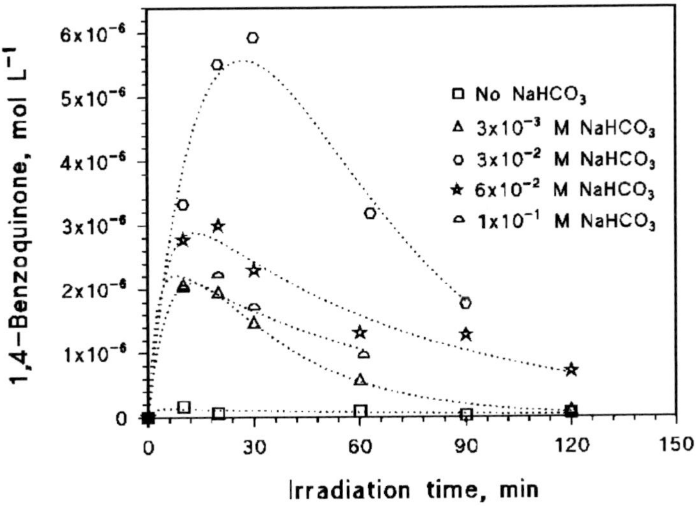

Figure 5. Time evolution of 1,4-benzoquinone upon UV-Vis irradiation of 0.1 mM phenol, 0.10 M 2-propanol, and 0.20 g L^{-1} TiO_2 (Degussa P25).

The formation of 1,4-benzoquinone is maximum at 0.03 M bicarbonate and it decreases when the concentration of $NaHCO_3$ is higher. A possible explanation is connected with the higher availability of $CO_3^{-\bullet}$ in the presence of $NaHCO_3$, as discussed before. The carbonate radical is well known to oxidise phenol to the phenoxy radical [19], which under photocatalytic conditions could follow two different pathways (see Scheme 1). One is the dimerisation to yield species such as 4-phenoxyphenol [32], which was not detected among the transformation intermediates of phenol. This finding suggests that the concentration of phenol used in the present study was not high enough, so as to allow the dimerisation process to compete effectively with the additional photocatalytic oxidation of the phenoxy radical. Phenoxyl could in fact be oxidised and yield hydroquinone upon water addition. Hydroquinone could then react with $CO_3^{-\bullet}$ and finally yield 1,4-benzoquinone [19] (Scheme 1). This process is in agreement with the experimental data, because the higher occurrence of $CO_3^{-\bullet}$ upon addition of $NaHCO_3$ would enhance the transformation of both phenol and hydroquinone and would yield a higher amount of benzoquinone.

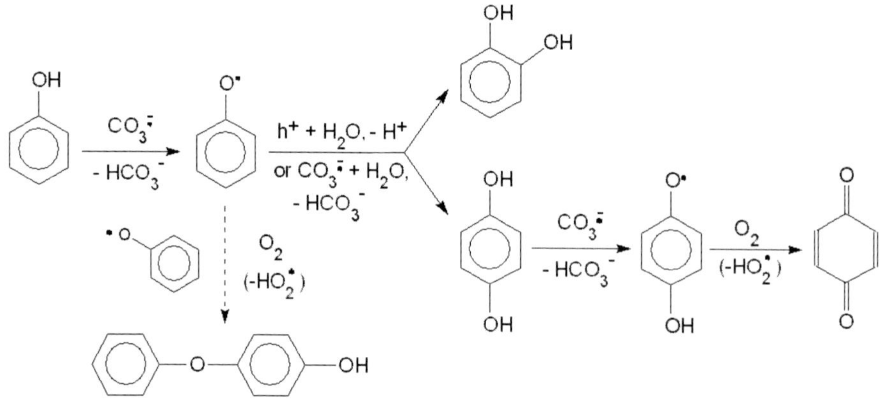

Scheme 1. Proposed early steps in the transformation pathways of phenol, in the presence of 2-propanol and $NaHCO_3$. The main pathways that are expected to occur in the presence of bicarbonate are presented here.

According to Scheme 1, the formation and the transformation of hydroquinone would both be enhanced by $CO_3^{-\bullet}$. This fact could account for the very variable trend of hydroquinone in the presence of different concentration values of $NaHCO_3$ (Figure 6).

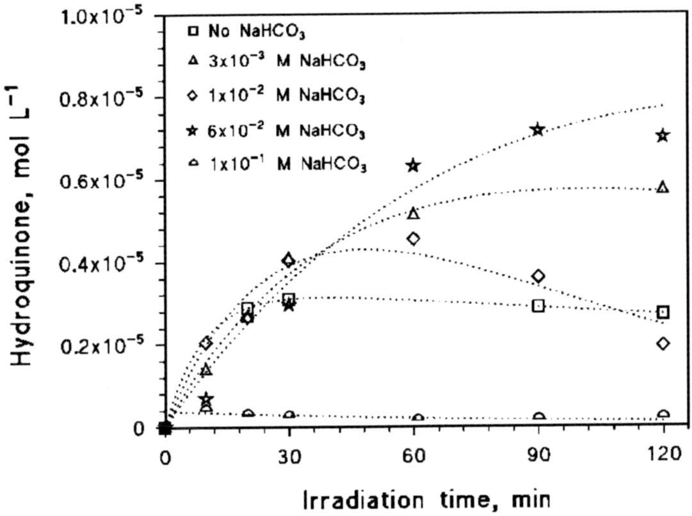

Figure 6. Time evolution of hydroquinone upon UV-Vis irradiation of 0.1 mM phenol, 0.10 M 2-propanol, and 0.20 g L^{-1} TiO_2 (Degussa P25).

Conclusion

In most systems of the kind substrate+$NaHCO_3$+TiO_2, bicarbonate and carbonate are well known to inhibit the photocatalytic degradation of the substrate. This work shows that the effect of bicarbonate may be quite different in four-component systems such as substrate+$NaHCO_3$+organic competitor+TiO_2, if the organic competitor reacts with $CO_3^{-\bullet}$ at a substantially lower rate than the substrate. In the presence of phenol as substrate and 2-propanol as competitor, $NaHCO_3$ enhances the degradation of phenol by shifting reactive species from the degradation of 2-propanol to that of phenol itself. This would happen because of the oxidation of carbonate and bicarbonate to $CO_3^{-\bullet}$, which shows selective reactivity toward phenol.

For the first time to our knowledge, this paper shows that the combined impact of bicarbonate and excess organic matter in photocatalysis may be very different from the sum of the effects of the separate components. The enhancement effect by bicarbonate was here observed at around 10 mM $NaHCO_3$, which is representative of waters with elevated alkalinity [33] and could, therefore, occur quite often in technological systems.

Acknowledgments

Financial support by Università di Torino - Ricerca Locale is kindly appreciated. The bursary of RD in Torino was financially supported by MIUR (Progetto India) and by Compagnia di San Paolo.

References

[1] Hoffman, M. R., Martin, S. T., Choi, W.&Bahnemann, D.W. (1995).*Chem. Rev*. 95, 69-96, and references therein.

[2] Brosillon, S.,Djelal, H.,Merienne, N.&Amrane, A. (2008).*Desalination*,222, 331-339.

[3] Saquib, M., Abu Tariq, M., Faisal, M.&Muneer, M. (2008). *Desalination*,219, 301-311.

[4] Ghaly, M. Y., Farah, J. Y.&Fathy, A. M. (2007). *Desalination*,217, 74-84.

[5] Fox, M. A.&Dulay, M. T. (1993).*Chem. Rev*.,93, 341-357, and references therein.

[6] Minero, C. (1999). *Catal. Today* 54, 205-216.

[7] Vione, D., Minero, C., Maurino, V., Carlotti, M. E., Picatonotto, T.& Pelizzetti, E. (2005).*Appl. Catal. B: Environ*. 58, 79-88.

[8] Minero, C.&Vione, D. (2006). *Appl. Catal. B: Environ*. 67, 257-279.

[9] Bhatkhande, D. S.,Pangarkar, V. G.&Beenackers, A. A. C. M. (2002). *J. Chem. Technol. Biotechnol*. 77, 102-116.

[10] Xu, Z. H., Jing, C. Y., Li, F. S.&Meng, X. G. (2008). *Environ. Sci. Technol*. 42, 2349-2354.

[11] Rincon, A. G.&Pulgarin, C. (2004). *Appl. Catal. B: Environ*. 51, 283-302.

[12] Christensen, P. A., Curtis, T. P.,Egerton, T. A., Kosa, S. A. M.&Tinlin, J. R. (2003). *Appl. Catal. B: Environ*.,41, 371-386.

[13] Wang, Y. B., Hong, C. S.& Fang, F. (1999). *Environ. Eng. Sci*. 16, 433-440.

[14] Djebbar, K.&Sehili, T. (1998). *Pest. Sci*. 54, 269-276.

[15] Jones, A. P.& Watts, R. J. (1997). *J. Environ. Eng. – ASCE*, 123, 974-981.

[16] Bekbolet, M.&Balcioglu, I. (1996). *Wat. Sci. Technol*. 34, 73-80.

[17] Zhu, X. D., Nanny, M. A.& Butler, E. C. (2007). *J. Photochem. Photobiol. A: Chem*. 185, 289-294.

[18] Canonica, S., Kohn, T., Mac, M., Real, F. J., Wirz, J.& Von Gunten, U. (2005). *Environ. Sci. Technol*. 39, 9182-9188.

[19] Neta, P.,Huie, R. E.& Ross, A. B. (1988). *J. Phys. Chem. Ref. Data*, 17 1027-1230.

[20] Hu, L. H., Flanders, P. M., Miller, P. L.&Strathmann, T. J. (2007). *Wat. Res.* 41, 2612-2626.

[21] Picatonotto, T., Vione, D.& Carlotti, M. E. (2002). *J. Disp. Sci. Technol.* 23, 845-852.

[22] Vione, D.& Carlotti, M. E. (2005). *J. Disp. Sci. Technol.*, 26, 163-168.

[23] Mazellier, P.,Busset, C., Delmont, A.& De Laat, J. (2007). *Wat. Res.* 41, 4585-4594.

[24] Larson, R. A.&Zepp, R. G. (1988). *Environ. Toxicol. Chem.*, 7, 265-274.

[25] Mrowetz, M.&Selli, E. (2005). *Phys. Chem. Chem. Phys.*,7, 1100-1102.

[26] Weber, M.&Kleine-Boymann, M. (2008).*Ullmann's Encyclopedia of Industrial Chemistry*, John Wiley and Sons, NY.

[27] Lathasree, S.,NageswaraRao, A.,Sivasankar, B.,Sadasivam, V.&Rengaraj, K. (2004). *J. Mol. Catal. A: Chem.* 223, 101-105.

[28] Robert, D.&Malato, S. (2002). *Sci. Tot. Environ.* 291, 85-97.

[29] Bekkouche, S.,Bouhelassa, M., Hadj Salah, N.&Meghlaoj, F. Z. (2004). *Desalination*, 166, 355-362.

[30] Minero, C., Mariella, G., Maurino, V., Vione, D.& Pelizzetti, E. (2000). *Langmuir*, 16, 8964-8972.

[31] Buxton, G. V.,Greenstock, C. L.,Helman, W. P.& Ross, A. B. (1988). *J. Phys. Chem. Ref. Data*,17, 513-886.

[32] Maurino, V., Borghesi, D., Vione, D.& Minero, C. (2008). *Photochem. Photobiol. Sci.*, 7, 321-327.

[33] Baker, A., Cumberland, S.& Hudson, N. (2008). *Area*,40, 117-127.

In: Photochemistry: New Research
Editors: A. B. Smith et al, pp. 37-76
ISBN 978-1-62808-615-7

Chapter 3

Conformational Relaxation of Excitons in Conjugated Polymers: A Stochastic Mechanical Model. Application to PPV (Poly-Phenylenevinylene) and P3HT (Poly-Thiophene) Oligomers

Adelino M. Galvão[1,*] *and K. Razi Naqvi*[2,†]
[1]Centro de Química Estrutural, Instituto Superior Técnico, Technical University of Lisbon, Lisbon, Portugal
[2]Department of Physics, Norwegian University of Science and Technology (NTNU), Trondheim, Norway

Abstract

Conjugated polymers draw considerable interest due to their electroluminescence and photovoltaic properties, and are seen as cheap alternatives to inorganic based optoelectronic devices. The underlying physics behind the properties of these compounds is rather complex and may involve solvent relaxation, intra or inter chain energy transfer and conformational relaxation of the polymer backbone. Since the exciton is not delocalized

*E-mail address: adelino@ist.utl.pt
†E-mail address: razi.naqvi@ntnu.no

over the entire chain but remains confined to just a few segments of the main backbone, the use of oligomers as model compounds enables us to disentangle several of the above mentioned photophysical processes. The problem of conformational relaxation in non polar solvents has been addressed by the authors (DOI: 10.1002/cphc.201200869) using a stochastic model based on a mechanochemical analog that includes the participation of the oligomer side chains in storing and dissipating the stresses induced by photoexcitation. The model was applied to a series of PPV (polyphenylenevinylene) trimers to predict the relaxation rate constants over a wide range of solvent viscosities as well as its temperature dependence. By treating transport properties as local quantities, the model allows prediction of viscosity in the nano-volume surrounding the oligomer.

In this work we propose to review in considerable detail the main features of our stochastic model extending the applications to new systems, in particular those which are predicted to relax in the sub pico-second timeframe such has PPVs without (or very small) side chains and/or P3HTs (3,4,4-trihexyl-2,2:5,2-terthiophene) poly-thiophenes which have a considerably smaller number of side chains then the previously reported systems.

Since non polar solvents such as chloroform are widely used to solubilize polymer films an extension of our viscosity model to polar solvents will also be addressed.

PACS 66.20-d, 83.60.Fg, 66.20.Cy, 33.50.Dq, 34.50.Ez, 78.47.D-.

Keywords: Conformational Relaxation, Fluorescence Decays, PPV oligomers, Viscosity, Energy Transfer

1. Introduction

Molecular materials depicting extended conjugated systems are becoming attractive as replacements for inorganic semiconductors in the manufacture of photo-voltaic and light-emitting devices as well as applications in optoelectronics [1–3], mainly due to the cheaper fabrication processes directly from solution. The performance of these devices depends on a myriad of processes, spanning from energy transfer to conformational relaxation of the molecular backbone, and a considerable amount of research has been devoted to the understanding of their photophysical behaviour [4–6]. Although the exciton may in a formal way delocalize over the entire backbone of a conjugated polymeric material, in fact, the conjugated length is limited to a few (2–12) monomeric units that act as chromophores in the absorption/emission processes [7, 8].

The fate of the exciton depends on the target application and while in a photovoltaic the aim is to separate the exciton into free charge carriers (in this context called polarons), in OLED's (Organic Light Emitting Devices) the recombination into a strongly bound exciton is the desired outcome. Either process usually doesn't happen instantaneously and require a series of sub events usually with different time scale kinetics. Tuning our device to a specific application requires a detailed knowledge of each of the processes, some of them in the femto and picosecond time frames. In particular, for the most popular conjugated compounds built with a framework of aromatic sub units that can librationally rotate relatively to each other in a nearly flat energy landscape around a twisted minimum position, the excitation stiffens the low frequency modes with an energy minimum around a flat geometry of the aromatic groups. The dynamics of the relaxation will depend on molecular characteristics of the molecule but in no lower extension on the environment variables such as temperature and viscosity. Except for the few examples where the excitation brings the molecule close to a non adiabatic seam that can energize the nuclei to reorganize in the sub-picosecond time frame [9], relaxation times in the picosecond (< 100 ps) window are common making pump-probe transient absorption spectroscopy and time-correlated single photon counting convenient methods for experimentally probing these processes [7, 10–13]. Apart from the above mentioned non adiabatic photoexcitations in the manifold of S_n, torsional relaxation is attended by vibrational cooling and the stretching of aromatic bonds by a few picometers (usualy within a single vibrational period of around 100 fs), and in solution is strongly opposed by the solvent. Maçanita *et al.* reported fast kinetics results for *p*-phenylenevinylene trimers, functionalized with terminal groups and side chains, in dilute solutions [10, 14, 15], showing that the relaxation rate constant depended heavily on solvent viscosity (η), temperature (T) and the van der Waals volume of the functional groups. The authors [16] have proposed a mechano-chemical stochastic model that explains the main features of conformational relaxation, in particular the unusual features observed in experimental Arrhenius plots of k_{CR} yielding activation energies (E_a) lower than the activation energy for viscous flow (E_η) and the non linearity observed in log-log plots of k_{CR} against viscosity η that can be as high as unity in low viscosity solvents and close to zero for high viscosity ones. Our aim here is to present a detailed description of our model, test it for oligomers without side chains and propose an extension of its application range to polar solvents. Figure 2 shows schematically the main features of our model:

i) The chromophore and its solvation layer (referred from now on as particle and host) constitute a single dynamical system;

ii) Photoexcitation brings the system to a non equilibrium torsional geometry. The force, induced by the energy gradient in the potential energy surface (PES), subjects the system to a shear stress; the shear stress is elastically linked to the shear strain of the solvent and it is assumed to follow Hooke's law with constant K; The shear strain can be described by a single conformational parameter χ, which can be visualized as the path of maximum energy gradient in the PES, linking the strained geometry and the minimum energy conformation, figure 1;

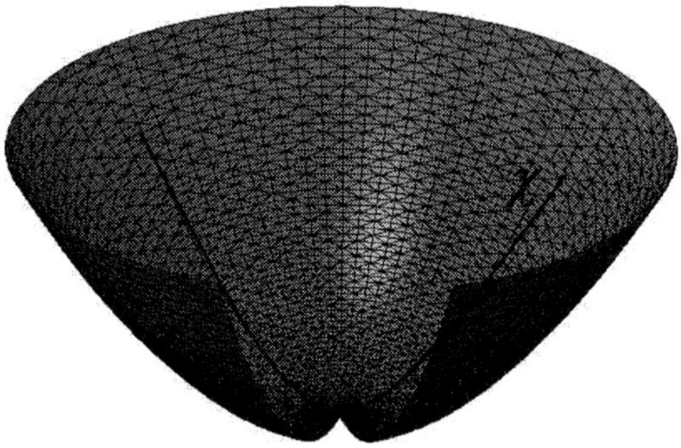

Figure 1. Maximum gradient path.

iii) The strain relaxes due to Brownian forces, $f(t)$, in the thermal bath, which follow Langevin dynamics governed by random gaussian white noise; The relaxation is mechanically modeled by storage (springs with constant G) and dissipative (dash pots with relaxation time τ) elements.

2. Mechanical Model

In the field of rheology, it has become customary to simulate the behavior of viscoelastic materials in terms of discrete mechanical components. Although molecules in a liquid or solid do not behave exactly like the individual elements

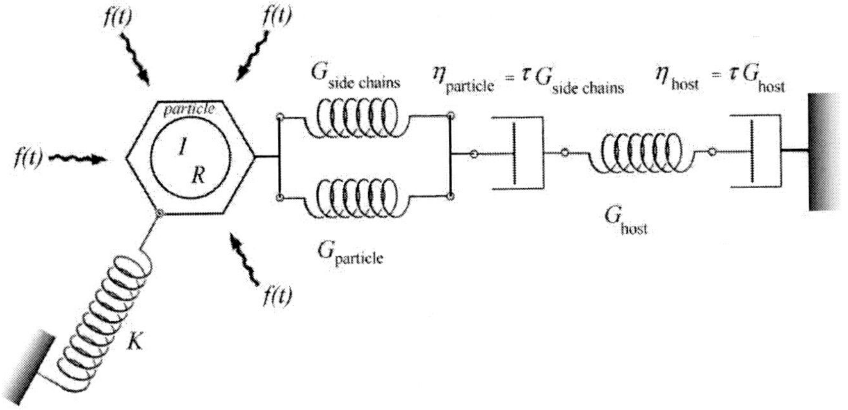

Figure 2. Mechanical analog of the molecular dynamic system. *I* and *R* are the moment of inertia and the characteristic length (radius of gyration) of the torsion spring.

of the models, modeling provides a framework for understanding and analysis of the behavior of viscoelastic materials. The main aims of such simulations, summarized by Crawford [17, pp. 84–103], are to facilitate the analysis of the behaviour of plastic products, to provide the possibility of extrapolating and interpolating experimental data, and to reducing the need for extensive, time-consuming creep tests. Some of the more important components and their combinations are presented below; for further details, the interested reader is referred to appendix and Crawford's textbook.

i) Springs: Modeled with a linear constitutive relation between stress and strain with the proportionality constants *K* or *G* representing, respectively, the stiffness of the spring or the shear modulus at infinite frequency of the system. When a stress is applied they strain instantaneously and hold the strain until the stress is removed.

ii) Dash pots: Depict a linear constitutive relationship between stress and strain rate with viscosity η as the proportionality constant. Under constant stress the strain builds up linearly with time at a rate that is proportional to the ratio between stress and viscosity (stress/viscosity).

iii) The series association of theses two elements creates a Maxwell model. The applied stress is felt equally by both elements but each one has its own share of the deformation. After the application of the stress ($t = 0$) the spring absorbs all the strain, but then ($t > 0$) there is an irreversible build up of strain in the dash pot at the rate defined in *ii*). If at a time t we remove the stress only the strain that remains in the spring can be recovered.

iv) The parallel association of a spring and a dash pot creates a Kelvin-Voigt model. The strain is the same in both elements but each one feels a different stress. After the application of the stress ($t = 0$) the dash pot takes all the stress, but as strain builds up ($t > 0$) there is a transfer of stress to the spring. This means there is a retardation time between the application of a stress and the development of a strain. If the stress is removed the dash pot will held the strain but with time the stress stored in the spring will bring the system to the original unstrained configuration.

v) More realistic models can be created by combining the four elements above: Combination of *i*) with one of the later two creates Standard solid models while the combination with *ii*) is appropriate for standard fluids. As an example the association of a spring in parallel with a Maxwell system originates the Zener model. The extra spring guarantees that after removal of the stress the Maxwell system will recover to the zero strain eliminating the unrealistic plastic deformation of elastic materials, figure 3.

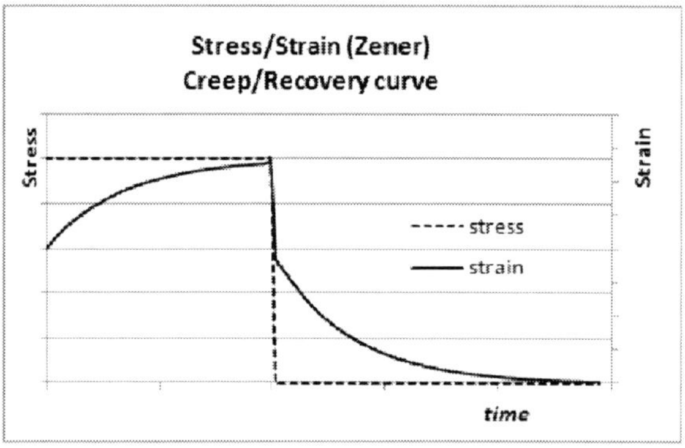

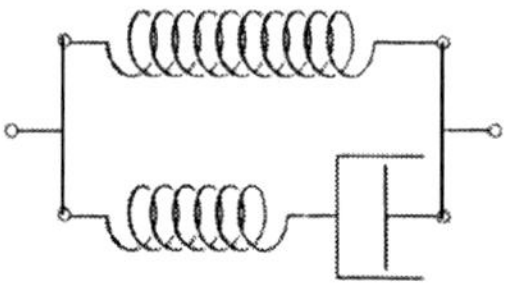

Figure 3. Creep/Recovery response for the Zener Model.

The mechanical elements just described will be used to construct our dynamical system under the following assumptions:

□ Any shear stress arising from geometric deformations due to photoexcitation will be stored by compression/expansion of the van der Waals interaction with the solvation layer (G_{particle}) in series with the similar linkage between solvation layer and thermal bath (G_{host}).

□ The presence of side chains that can act as entropic springs will add a shear relaxation modulus ($G_{\text{side chains}}$) in parallel with (G_{particle}).

□ The combined effect of van der Waals interaction and side chains will be represented by a single spring (G_{part}).

□ Both the host and the particle are subjected to gaussian white noises, $f(t)$, described by:

$$\begin{aligned} \langle f_{\text{part}}(t)\rangle &= 0 = \langle f_{\text{host}}(t)\rangle \\ \langle f_{\text{part}}(s) f_{\text{part}}(t)\rangle &= 2k_B T \gamma_{\text{part}} \delta(s-t) \\ \langle f_{\text{host}}(s) f_{\text{host}}(t)\rangle &= 2k_B T \gamma_{\text{host}} \delta(s-t) \end{aligned}$$

where k_B, T, γ's and δ are, respectively the Boltzmann constant, the temperature, the friction coefficients and the delta function. Quantities between angular brackets are ensemble averages.

Under no-slip conditions (sticky solvation layer) and torsional shear the hydrodynamics of the system is similar to that of a rotating sphere with the viscosity related to the friction coefficients by $\gamma = 8\pi\eta R^3$ [18]. This relaxation dynamics is introduced in the mechanical model through dash pot elements η_{part} and η_{host}. The net effect of the several mechanical elements represented in figure 2 can be reduced by reducing it to a Zener equivalent of figure 3, by use of equations (1) and (2), figure 4:

$$G = \frac{G_{\text{host}}\, G_{\text{part}}}{(G_{\text{host}} + G_{\text{part}})}, \tag{1}$$

$$\eta = \frac{\eta_{\text{host}}\, \eta_{\text{part}}}{(\eta_{\text{host}} + \eta_{\text{part}})}. \tag{2}$$

A contestable point in the mechanical representation of the dynamic system in figure 2, which translates in equation (2), is the use of a common relaxation time τ for both dissipators: the particle and the host. However this problem has been addressed by Magee in 1974 [19] by pointing out that in steady state the torque in the particle and host must be equal generating a local apparent friction coefficient γ:

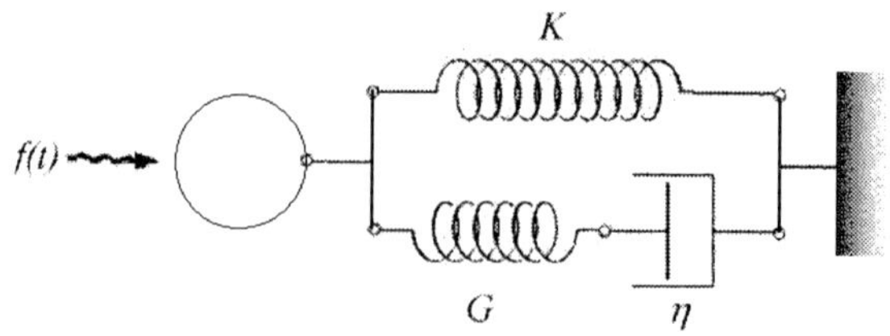

Figure 4. Zener model of dynamical system.

$$\tau = \frac{\gamma}{2KT}$$
$$\text{torque} = \gamma\dot{\chi} = \gamma_{\text{part}}\dot{\chi}_{\text{part}} = \gamma_{\text{host}}\dot{\chi}_{\text{host}}$$
$$\dot{\chi} = \dot{\chi}_{\text{part}} + \dot{\chi}_{\text{host}}$$
$$\tau = \frac{1}{2KT}\frac{\gamma_{\text{host}}\gamma_{\text{part}}}{\gamma_{\text{host}} + \gamma_{\text{part}}}$$

Comparing the system of figure 4 with the creep/recovery curve of figure 3 we can qualitatively predict how our system responds to the stresses induced by photoexcitation: The excited state system with a parabolic PES projection on the reaction coordinate χ is stressed, within the time frame of the electronic transition, by a force $K\chi$ which translates in an instantaneous strain χ. This strain is held as solvent shear and, depending on the number and size of the side chains, as conformational energy. Due to the overdamped conditions and stiffness of the potential any librational movement along the reaction coordinate is quickly arrested and the recovery to zero strain follows the exponential profile of figure 3. Equation (2) allow us to predict that in low viscosity solvents dissipation by the side chains is negligible while in the opposite end of the viscosity range the side chains are the only available means for strain relaxation.

This qualitative approach, to be applied to experimental data, has still to be dressed with appropriate tools to compute the characteristic relaxation time, which can not be achieved without relating the parameters of our dynamic system K, G and η with properties of solvent and solute.

3. Stochastic Model

The particle in figure 4, characterized by the moment of inertia I, is subjected to the action of deterministic forces and stochastic forces. These forces produce torques with a resultant moment responsible for the tangential acceleration $R^{-1}d^2\chi/dt^2$. Except for the perturbation stress $(K\chi/R)$ all the other moments are a consequence of the white noise (visualized as collisions of solvent molecules with the particle) which by virtue of happening in an interval very much shorter than the experimental time resolution (0.1–1 ps) will behave as gaussian.[1] Since the stress and strain are harmonically bound, the relaxation of one will induce a proportional relaxation of the other, creating moments contributing to the overall torque acting on the particle: stress $((\tau K/R)d\chi/dt)$ and strain $((\gamma/R)d\chi/dt)$ relaxation. The fluctuating force $f(t)$ is the only external force in the sense that it does not depend on the strain parameter χ or any of its derivatives. The couple introduced by the external random force depends on the geometry of collisions and only its tangential component is effective to the dynamics of our particle. The inelastic nature of solvent-solute collisions introduces the necessity of the particle to store the excess energy in other normal modes and an effective couple $R\, f_{\text{eff}}(t)$ will be used in the calculations.

The equilibrium of moments in figure 4 requires the balance between internal and external moments resulting in the Langevin equation (3).

$$R\, f_{\text{eff}}(t) = I\,\frac{d^2\chi}{dt^2} + K\,\chi + \gamma\,\frac{d\chi}{dt} + \tau\,K\,\frac{d\chi}{dt}, \tag{3}$$

where τ is obtained by the Maxwell relation $\tau = \eta/G_\infty$. G_∞ stands for the shear modulus at infinite frequency and experimentally can be identified with the high-frequency plateau value of the modulus [20]; in the remaining of the paper we will make no further distinction with G dropping the subscript ∞.

If we write equation (3) in a more practical form in terms of the acceleration:

$$\frac{d\dot{\chi}}{dt} = -\,\beta\,\dot{\chi} - \omega^2\chi + A(t) \tag{4}$$

where $A(t)$ is a random variable that satisfy a fluctuation-dissipation theorem [21] of the second kind:

[1] This is a consequence of the central limit theorem. The observable noise is a convolution (mean) of many statistical events of arbitrary probability distribution and is normally distributed.

$$\langle A(t)\, A(s)\rangle = 2\,\frac{k_{\mathrm{B}}T}{I}\,\beta\,\delta(s-t),$$

with

$$A = \frac{R^2}{I} f_{\mathrm{eff}}(t), \qquad \beta = \frac{\gamma}{I}\left(1 + \frac{1}{8\pi R^3}\frac{K}{G}\right) \qquad \text{and} \qquad \omega = \sqrt{\frac{K}{I}},$$

we get equation (4), reducing our problem to the Brownian motion of an harmonically bound particle. Equation (4) was formally introduced by Uhlenbeck and Ornstein [22] as a case study of a particle bound by a generic potential but an analytic solution was obtained by Ornstein and van Wijk [23, 24]. In the high friction limit one can ignore the inertia term of equation (4), and write

$$\beta\,\dot{\chi} = -\,\omega^2\chi + A(t), \tag{5}$$

and one is led [25] to equation (6), established first by Smoluchowski [26]:

$$\frac{\partial P(\chi,t)}{\partial t} = \frac{k_{\mathrm{B}}T}{\beta}\frac{\partial^2}{\partial\chi^2}P(x,t) + \frac{\omega^2}{\beta}\frac{\partial}{\partial\chi}[\chi\, P(x,t))]. \tag{6}$$

When the dynamic system is perturbed by a sudden strain $P(\chi,0) = \delta(\chi - \chi_0)$, the fundamental solution is a gaussian whose mean value and variance are both time-dependent:

$$P(\chi,t) = \frac{1}{\sqrt{2\pi\sigma^2(t)}}\exp\left[-\frac{\left(\chi - \chi_0 e^{-kt}\right)^2}{2\sigma^2(t)}\right], \tag{7}$$

where

$$k = \frac{\omega^2}{\beta} \quad \text{and} \quad \sigma^2(t) = \frac{k_{\mathrm{B}}T}{\beta k}\left(1 - e^{-2kt}\right).$$

Equation (7) represents a gaussian distribution which recovers to the zero strained conformation with a relaxation time $\tau = \beta/\omega^2$, equation (8):

$$\tau = \frac{\beta}{\omega^2} = \eta\left(\frac{8\pi R^3}{K} + \frac{1}{G}\right) = \eta\left(\frac{1}{K'} + \frac{1}{G}\right), \tag{8}$$

Equation (8) depends only on three parameters, G, η and K'. The first two can be calculated from equations (1) and (2) provided we can estimate the local properties of the solution in the nanovolume of the structured particle with its sticky solvation layer. K' can be estimated by topological analysis of the excited state PES (Hessian at ground state geometry) or adjusted to experimental relaxation times [16].

The following sections will propose approaches based on QSAR (Quantitative Structure Activity Relationships) for estimating G and η and applying them to the prediction of relaxation times in case studies.

4. Viscosity Model

The assumptions made to develop our stochastic model parallel those of the Dyre's shoving model [27]: Our particle is surrounded by a sticky layer of solvent elastically bound to both the particle and the surrounding thermal bath; The elastic energy generated by structural rearrangements is stored as shear without any macroscopic change of volume; and finally this elastic energy is the main contribution to the activation free energy. According to Dyre and Wang [20, 27], under the above assumptions the free energy can be calculated by:

$$\Delta^{\sharp}G = G\,V_{\mathrm{c}}, \quad \text{with} \quad V_{\mathrm{c}} = \frac{2}{3}\frac{\left(V_{*} - V_{\mathrm{vdW}}\right)^{2}}{V_{\mathrm{vdW}}}F. \tag{9}$$

In equation 9, V_{c} is a "characteristic" microscopic volume (see figure 5), V_{vdW} is the van der Waals volume of the particle, V_{*} is the volume available per particle and F is related to the asphericity of the particle and will be defined elsewhere.

Introducing Dyre's definition of activation free energy into the viscosity equation of Ewell and Eyring [28], we can relate the "local" viscosity to the shear modulus, already present in our mechanical model, and molecular volumes, easily obtained from densities and molecular structures. Whence follows the relation

$$\eta = \frac{h}{V_{\mathrm{c}}}\exp\left(\frac{\Delta^{\sharp}G}{k_{\mathrm{B}}T}\right) = A_{\mathrm{int}}\frac{h}{V_{\mathrm{c}}}\exp\left(G\,\frac{V_{\mathrm{c}}}{k_{\mathrm{B}}T}\right). \tag{10}$$

The parameter A_{int}, takes the burden of all the internal modes contribution to entropic terms in the activation energy. The calculation of A_{int} can be made, minus a global constant, from the ratio of the internal modes partition function at rest and the same function without the internal mode involved in the flow [28]. Applying the concept to translational motion would give us [29], equation (11):

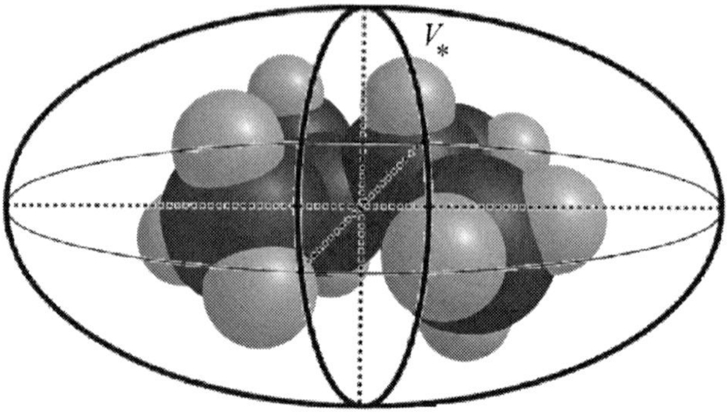

Figure 5. Volume per particle.

$$A_{\text{int}} = c\frac{(2\,\pi\,M\,k_{\text{B}}T)^{1/2}}{h}\left(V_*^{\frac{1}{3}} - V_{\text{vdW}}^{\frac{1}{3}}\right). \tag{11}$$

In equation 9, M is the mass of one molecule and c is a constant applied to all systems. It is worth noting that our mechanical scheme, figure 2, includes a $G_{\text{side chains}}$ (entropic spring) in series with G_{particle} (elastic spring) and as so a further temperature independent contribution to the pre-exponential factor of viscosity will appear from the term $\exp\left(G_{\text{side chains}}V_{\text{c}}/k_{\text{B}}T\right)$, since entropic shear modulus are proportional to $k_{\text{B}}T$. The only parameter that remains to be estimated is the shear modulus G. In the framework of Dyre's shoving model [20, 27] we can estimate the value of the shear modulus, for pure solvents, from experimental viscosity data, extracting activation energies from Arrhenius plots and computing G as the ratio between this energy and the "characteristic" volume V_{c}, equation (9).

Unfortunately, it is technically impossible to obtain experimental data for our dynamic system (with a nano dimension), where a particle interacts with a chemically different host (solute and solvent). Even for pure substances data is not always available and, to fill the gap, the next subsections will deal with the development of a QSAR approach, based on molecular parameters, to estimate local viscosities, η, and shear modulus, G.

4.1. Elastic Shear Modulus of Pure Substances (G_{host})

As described above,when a liquid is suddenly perturbed by a force it behaves as a solid and the response is controlled by the shear modulus G and the modulus of compression. Zwanzig and Mountain [30] showed that for an isotropic media the two quantities are related by a factor of $5/3$ and that the later could be computed by statistical mechanics from the radial distribution function (RDF) if the interacting potential is known. Developing a QSAR approach does not require the explicit analytical solution of the RDF for the chosen potential but only the knowledge of the key quantities that capture the main features of it. Since our model assumes a harmonic binding between particle and host the realistic Lennard-Jones potential will be series expanded up to the second term, on the assumption that the stress induced by photoexcitation is small enough to avoid the anharmonic region of strain, equation (12) and figure 6

$$V = \varepsilon\left[\left(\frac{\sigma_0}{r}\right)^{12} - 2\left(\frac{\sigma_0}{r}\right)^{6}\right] = -\varepsilon + \frac{36\,\varepsilon}{{\sigma_0}^2}(r - \sigma_0)^2 + o\left((r-\sigma_0)^4\right) \quad (12)$$

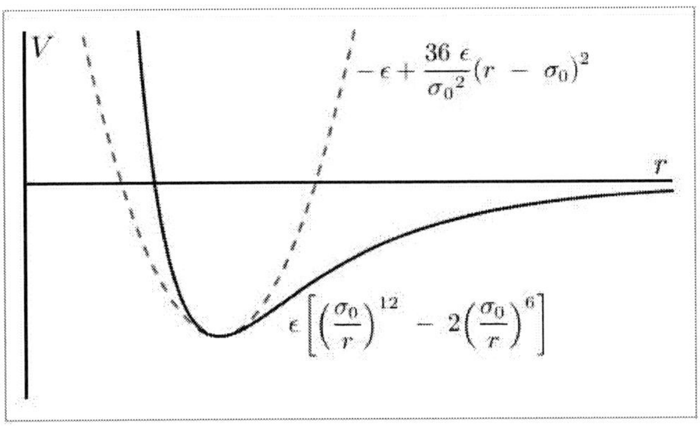

Figure 6. Harmonic expansion of LJ potential.

The parameter ε is the depth of the potential well, σ_0 is by construction the distance at which the potential depth is calculated and $\sigma = \sigma_0/2^{1/6}$ is the van der Waals diameter, equation (12).

In our previous work [16], ε was estimated from the C_6 coefficient of the London dispersion forces with a polarizability proportional to:

$$\frac{(FV_{\text{vdW}})/n^{1/2}}{E_{\text{i}} - E_{\text{j}}},$$

(FV_{vdW}) is the spherically equivalent van der Waals volume and

$$G \propto \frac{(FV_{\text{vdW}})^2/n}{\sigma^2\left(E_{\text{i}} - E_{\text{j}}\right)}, \tag{13}$$

where $E_{\text{i}} - E_{\text{j}}$ is taken to be the ionization potential (or, if there is a low lying electronic transition, the geometric mean between this energy and the ionization potential [31]). This correlation was found to predict good values of viscous activation energies for linear alkanes, toluene and methylcyclohexane when compared to those derived from Arrhenius plots.

In this work we propose to test the transferability of the proportionality constant previously determined, by applying it to the 6-carbon series of branched alkanes, figure 7:

- □ 2-Methylpentane (2MP)
- □ 3-Methylpentane (3MP)
- □ 2,2-Dimethylbutane (22DMB)
- □ 2,3-Dimethylbutane (23DMB)

Although the application of the procedure based on the C_6 coefficient seems to produce reasonable results, it will be harder to apply to polar solvents because it would require the explicit introduction of the interaction between the dipole moment of the host with the induced/instantaneous dipoles of the particle. A new formalism, based on Oliveira [34], will be introduced, which considers the interaction between a finite object and a surface, placed at half the interlayer distance σ_0. Hamaker [35] developed an expression for the attractive part of the vdW potential per unit area, proportional to the C_3 coefficient and inversely proportional to the cube of distance to the reference surface. By differentiation and making the C_3 coefficient proportional to FV_*/I_p, the polarizability function, (I_p is the ionization potential of the particle) and $(\varepsilon - 1)/(\varepsilon + 1)$, the high frequency dielectric function, we obtain a relation, equation (14), not dependent on microscopic structural parameters:

$$G \propto \frac{(FV_*)(FV_*)^{2/3}}{\sigma^4\, I_p}\,\frac{\varepsilon - 1}{\varepsilon + 1}, \tag{14}$$

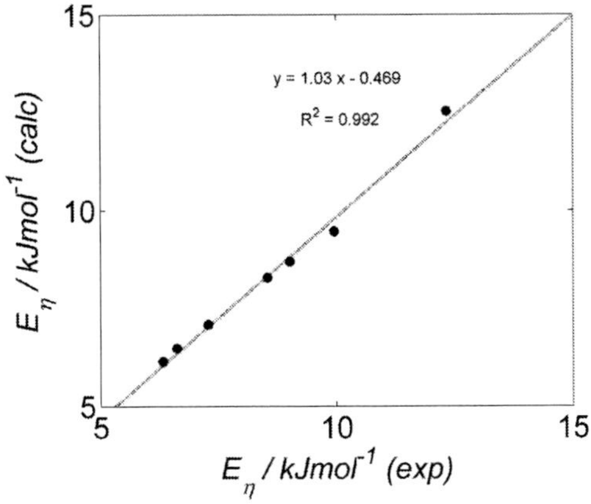

Figure 7. E_η calculated from GV_c versus E_η from Arrhenius plot for the 6-carbon branched alkanes, butane, pentane and hexane. (Arrhenius plots based on data compiled by C. Wohlfarth and B. Wohlfarth [32] and Iwahashi and Kasahara [33]).

where $(FV_*)^{2/3}$ is the spherically averaged area of the cavity,and $\varepsilon = n_D^2$ and n_D the refractive index.

Equation (14) explicitly includes the dielectric constant enabling us to extend it to the treatment of polar solvents. The effect of the permanent dipoles is essentially felt by the center of masses and does not contribute to the fixation of the solvation layer under a shear stress. Nonetheless dipoles create electric fields that are going to be felt by the neighborhood molecules. The external field will enhance the dielectric response and its contribution can be evaluated by the Guoy-Chapman model [36] using the thin double layer model applied to a spherical particle [37]. The electric field created by the dipoles in the host decays exponentially from the solvation layer to the surface of the particle. If the polarization potential is large enough the enhancement will diverge to the limiting value given by equation (15)

$$\varepsilon_S = \frac{9}{4}\,\varepsilon\,\varphi\,\exp\left[\frac{V_{\text{polar}}}{V_{\text{non polar}}}\right], \qquad (15)$$

in which ε_S is the static dielectric constant, φ is the fraction of occupied volume, $V_{\text{non polar}}$ is the potential felt in an equivalent non polar solvent (only dispersion forces) and V_{polar} is the potential actually felt by the system. Since ε_S is experimentally available we can use equation (15) reversivelly to obtain an enhancement factor to the potential which therefore has to be applied to the shear modulus G, equation (16).

$$G \propto \frac{(FV_*)(FV_*)^{2/3}}{\sigma^4\, I_p} \frac{\varepsilon - 1}{\varepsilon + 1} \ln\left(\frac{\varepsilon_S}{\varepsilon}\right). \tag{16}$$

The model was applied to the n-alkane series (n = 4–17), the 6-carbon branched alkanes, methylcyclohexane, toluene, tetrahydrofuran (THF), dimethyl formamide (DMF) and dimethyl sulphoxide (DMSO) at 20°C, figure 8.

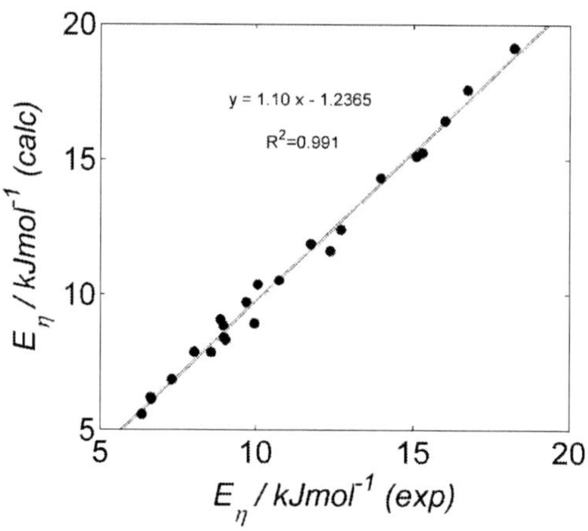

Figure 8. E_η calculated from GV_c versus E_η from Arrhenius plots based on data compiled by C. Wohlfarth and B. Wohlfarth [32] and Iwahashi and Kasahara [33].

4.2. "Entropic" Shear Modulus of Carbon Chains ($G_{\text{side chains}}$)

Our starting point to calculate the entropic contribution to viscosity is the free ideal chain with n segments of length d and a probability distribution of end-to-end distances (R_d) given by equation (17):

$$P(R_d, n) = \left(\frac{3}{2\pi n d^2}\right)^{3/2} \exp\left(\frac{-3R_d^2}{2nd^2}\right), \tag{17}$$

Knowing that in the activated state of viscosity the end-to-end distance should be as close to zero as possible (making the exponential term negligible) and in the relaxed state should behave as an expanded chain, the ratio of conformational partition functions can be predicted to follow the ansatz, equation (18):

$$A_{\text{chains}} = (n/2)^{3a\varphi} \exp(-\varphi a n/2), \tag{18}$$

where φ is the fraction of occupied volume and *a* is a proportionality constant. Figure 9 compares the pre-exponential factors predicted by our model and experimental values calculated from experimental viscosities and activation energies previously calculated from Arrehnius plots.

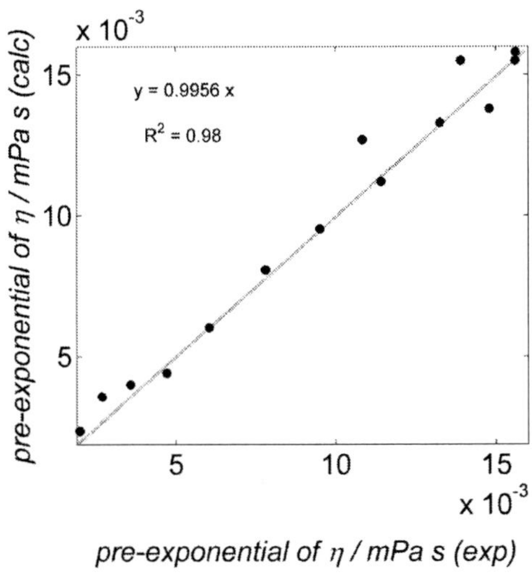

Figure 9. Calculated versus experimental pre-exponential factors of viscosity ($a = 1.594$). Experimental pre-exponential factors calculated from Arrhenius plots based on data compiled by C. Wohlfarth and B. Wohlfarth [32].

A more analytic treatment of chains based on the Rouse model [38, 39] can be found in [16], equation (19), the last term accounts for corrections introduced using the Doi and Edwards model [40], corrected for contour length fluctuations (CLF), in accordance with Doi [41, 42].

$$\eta = \frac{\rho N_A k_B T}{M} \frac{\zeta n^2 d^2}{36 k_B T} + 3\tau_1 \left(\frac{M}{M_c}\right)^3 \left[1 - \kappa \left(\frac{M}{M_c}\right)^{\frac{1}{2}}\right]^3, \tag{19}$$

$$\tau_1 = \frac{\zeta d^2}{3 k_B T} \left(4 \sin^2 \frac{\pi}{2n}\right)^{-1} \tag{20}$$

In equation (19) n is the number of $-C-$ tetrahedral units in the chain, d is the distance between the vertices of the two half $C-C$ bonds attached to the carbon atom, ζ ($\zeta = 3.62$ Npsm^{-1} [16]) is the friction coefficient of a single $-C-$ unit, ρ is the mass density, N_A is the Avogadro number, M the molecular weight, M_c is the molecular weight of a Kuhn segment (optimized as eight $-C-$ units) and κ a best fit parameter ($\kappa = 0.442$, [16]). In equation (20) τ_1 is the longest Rouse relaxation time.

4.3. Shear Modulus in Multicomponent Solutions

Photochemical studies are usually performed in dilute solutions, where a macroscopic homogeneity of properties is observed but, at the nanoscale in a volume around a solute particle, the properties are different from the bulk of the solvent. Furthermore, to achieve certain combinations of viscosity and temperature, complex solvent mixtures are regularly used. Solvent mixtures were previously treated [16] by replacing the mixture by a single molecule with the same properties (viscosity, activation energy) measured for the mixture. The procedure involves the use of our viscosity model, sections 4.1 and 4.2, to design a molecule that matches experimental properties as well as having the right balance between enthalpic and entropic shear modulus, usually achieved by using a model compound of the same chemical class. Within our present formulation of interaction between a particle and the surface of the host all we need is an extra experimental measurement of the dielectric constant at low and high frequencies, the former being dispensable for non polar solvents and the later is just a measurement of the refractive index.

The computation of C_6 parameters between unlike molecules or C_3 reference surface placement in particle host interactions, will recour to a set of combining rules such as the Lorentz-Berthelot:

$$C_3 \quad \text{or} \quad C_6 \quad \sigma_{ab} = \frac{(\sigma_a + \sigma_b)}{2} \tag{21}$$

$$C_6 \quad \varepsilon_{ab} = \lambda(\varepsilon_a \varepsilon_b)^{1/2}; \qquad \lambda < 1$$

$$C_6 \quad I_{ab} = \frac{2\, I_a\, I_b}{(I_a + I_b)}; \qquad I_x = (E_i - E_j)_x; \quad x = a,b.$$

The parameter λ takes account of deviations from the Berthelot geometric mean rule in the direction of lower cohesion of the a-b pair.

5. Shear Stress K'

When a vast array of experimental data such as relaxation times as a function of viscosity and/or temperature are available K' can be calculated by a least squares analysis of experimental and modeled relaxation times. Unfortunately, data are often available only for one or two solvents, and usually for a single temperature. To apply our model in such circumstances we can have recourse to quantum chemical methods for finding an estimate that can be fed to our model. In this section we will apply our suggested procedure to the simple oligomer of PPV (poly-phenylenevinylene) (E)-2,2'-dimethylstilbene.

We start our procedure by optimizing the molecular structure of the ground state using one the available methods such as HF/6-31++G** included in GAMESS [43, 44]. The second step is to obtain an Hessian, with full vibrational analysis, computed for the relevant excited state (S_1). Our procedure used a semi-numerical approach at CIS (Configuration Interaction Singles) level producing an imaginary frequency of 167.95 cm^{-1} with a reduced mass of 3.2059 atomic units. Figure 10 shows the geometry and vectors of the "imaginary"[2] normal mode. K and K' can then be computed by equation (22),

$$K = \frac{\nu^2\, \mu\, R}{1000 N_A S} \qquad K' = \frac{K}{8\pi R^3}, \tag{22}$$

where ν is the imaginary frequency, μ is the reduced mass, N_A the Avogadro number, R the radius of an equivalent spherical volume and S the surface area of the molecule. The computed values are $K = 19$ GNm and $K' = 14.3$ MPa.

[2] The frequency was used only to compute the force constant of the only significantly non zero gradient which is nearly tangent to a gradient extremal. The IRC roughly follows this extremal.

Previously reported optimized values (based in experimental data) for trimers [16] fall in the range 16–66 MPa.

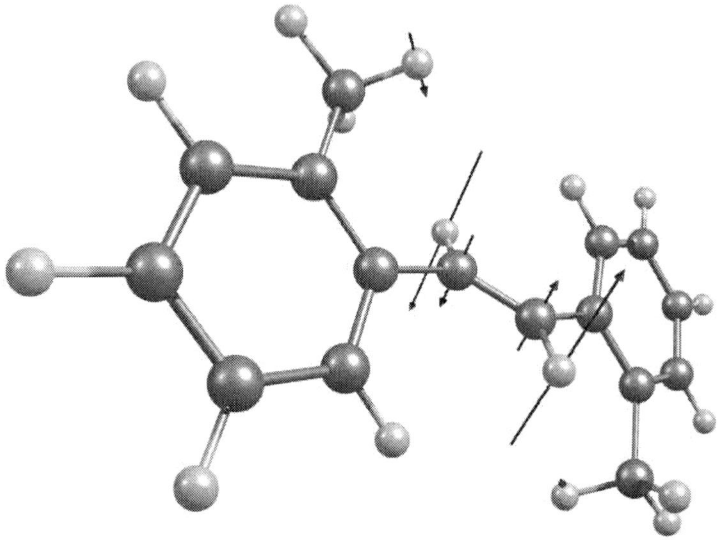

Figure 10. Imaginary-frequency normal mode.

It is interesting to note that an imaginary frequency means that we are above the valley-ridge crossing point and that the equation of motion will, at a certain point, change from a relaxation exponential decay to a damped oscillatory motion. In fact if we define a complex variable $z = \chi + iy$ and a complex frequency $\Omega = \kappa + i\,\omega$, and consider the equation of motion

$$\frac{dz}{dt} = -\,\Omega\,\chi,$$

we will find, after separating the real and imaginary parts, that χ and y satisfy two first-order coupled equations, and both satisfy the same second-order differential equation, namely $[D^2 + 2kD + (k^2 + \omega^2)]u = 0$, where $u = \chi$ or y, and $D^n \equiv d^n/dx^n$. The solution, with $u = \chi$ may be expressed as [45]:

$$\chi = \text{constant} \times \exp(-\,\kappa\,t)\cos(\omega\,t + \phi). \tag{23}$$

Equation (23) shows that relaxation ($\omega = 0$), undamped oscillatory motion ($\kappa = 0$) and underdamped oscillatory motion ($\omega \neq 0$ and $\kappa \neq 0$) can all be encapsulated within the same first order linear differential equation and coexist in the lifetime of the process. A relaxation can be seen as a vibration with an imaginary frequency; conversely, a vibration can be viewed as a relaxation with an imaginary relaxation rate.

6. Results and Discussion

Experimental relaxation times for conformational relaxation follow a $\eta_{\text{pure solvent}}^{-\alpha}$ dependency and it happens as an experimental observation that α is also the proportionality constant between the activation energy for conformational relaxation and the activation energy for viscous flow of the pure host. Equation (1) can be written as $G = \alpha G_{\text{host}}$ with $\alpha = G_{\text{part}}/(G_{\text{host}} + G_{\text{part}})$. Since we have to maintain hydrodynamic equilibrium a decrease by α in the shear modulus requires a compensatory increase in strain (with a linear constitutive relation with the fraction of occupied volume φ). Equation (10) can be expressed in terms of α as,

$$\eta_{\text{host}} = \left(h\, A_{\text{rigid}}\right)^{1-\alpha}\, \eta_{\text{pure solvent}}^{\alpha} \tag{24}$$

6.1. Comparison Between Experimental, C_6 Particle/Particle Potential Model and C_3 Particle/Surface Model for Calculated Relaxation Times

In our previous work [16] relaxation times were calculated using a C_6 particle/particle potential model. The new particle/surface potential proposed here is used to compare the results of both models with experimental values. Of particular importance is that the new model has one less parameter since we dont need to optimize the λ coefficient in the Lorentz-Berthelot combining rules to account for the lower cohesion of dislike particles. Figure 11 shows the molecular graph of MBOPV3 and table 1 compiles the relaxation times obtained experimentally and by each of the theoretical models. Table 2 collects data for a PPV trimer with longer and less branched side chains, figure 12. Amine-end-capped trimers (NP63 and NP68) with side chains similar to MBOPV3, figure 13, where also tested to account for the effect of the termination groups, specially if they introduce the possibility of extending the aromatic system, table 3. Table 4 summarizes the results optimized for K' and η_{particle}. Figure 14 represents the overall fitting for the three model compounds.

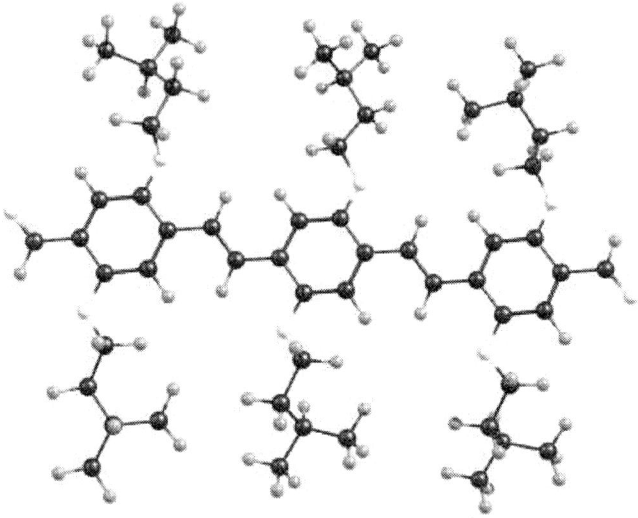

Figure 11. 2,5-di-[(3-methylbutyl)-oxyl]-p-phenylenevinylene (MBOPV3).

Table 4 captures the main essence of the factors controlling torsional relaxation in PPV trimers:

☐ The bulkier the side chains the greater the stress K'.

☐ The longer the side chains the greater the viscosity enhancement.

☐ Only side chains are strain relaxation enhancers. The end-capping groups just contribute to increase the elastic linkage to the host.

☐ In the event the end-capping groups, such us amines, can extend the aromatic system, the torque K and thereby the stress K' is substantially reduced.

☐ Increasing the size of the end groups just marginally decreases the stress K' essentially by spreading the torque K by a greater contact surface.

6.2. Model Extension to Systems with Single Sided Chains or Without Chains at All

The photophysics of PPV oligomers without side chains like DSBI, figure 15, have been studied by Maçanita and Seixas de Melo[3] and the results are not compatible with a solvent driven relaxation with values in the sub picosecond

[3]private communication of preliminary results

Table 1. Experimental and Calculated (from equation (8)) Values of Relaxation Times of MBOPV3 in Pure non Polar Solvents and *n*-Hexadecane:Liquid Paraffin Mixtures

trimer	solvent	$\tau(\exp)^a$ (ps)	$\tau^b_{model}(C_6)$ (ps)	$\tau_{model}(C_3)$ (ps)
MBOPV3	pentane	7	10	13
MBOPV3	hexane	12	13	15
MBOPV3	nonane	24	24	26
MBOPV3	decane	25	28	30
MBOPV3	undecane	36	32	33
MBOPV3	dodecane	41	36	37
MBOPV3	tetradecane	52	44	44
MBOPV3	hexadecane	52	50	50
MBOPV3	l. paraffin	65	68	68
MBOPV3	paraffin 90%	66	68	67
MBOPV3	paraffin 80%	68	66	66
MBOPV3	paraffin 70%	59	65	64
MBOPV3	paraffin 50%	63	61	61
MBOPV3	paraffin 25%	56	54	54
MBOPV3	toluene	18	24	22
MBOPV3	MCH	24	27	26

[a] from reference [14] [b] from reference [16]

time frame. According to our analysis the assumption of a sticky linkage between solute and host requires a zero shear rate at the surface interface. this condition can only be achieved if a flexible anchor (side chains) can be used as a “media” through which the gradient of the shear rate develops. The only anchors in DSBI that can prevent free rotation are the hydrogens. Consequently, when a stress is strong enough to overcome static friction the system changes regime from a sticky to slip condition (with a non zero shear rate at the interface with the host). To accommodate the change of regime within our model we will evolve it to the limit where the hydrogens have a minimal (close to, but greater than, zero) viscosity. Using the preliminary experimental results we were able to compute the minimal viscosity as 0.02 mPa s.

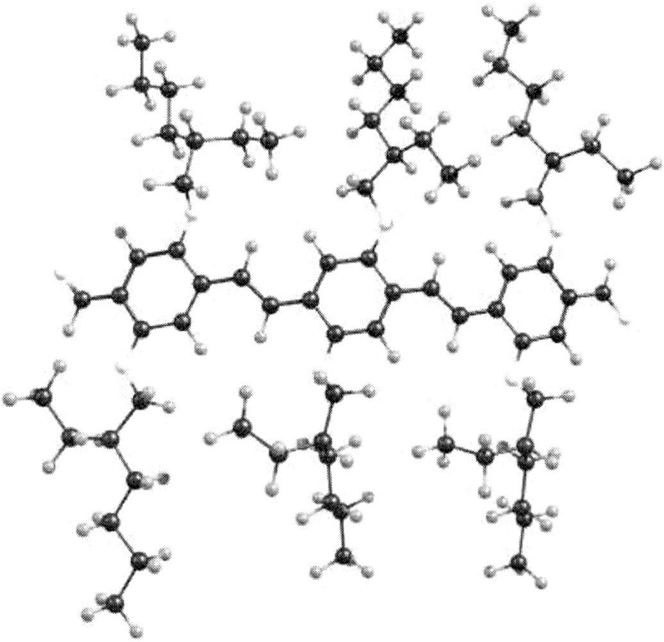

Figure 12. 2,5-di-[(2-ethylhexyl)-oxyl]-p-phenylenevinylene (EHOPV3).

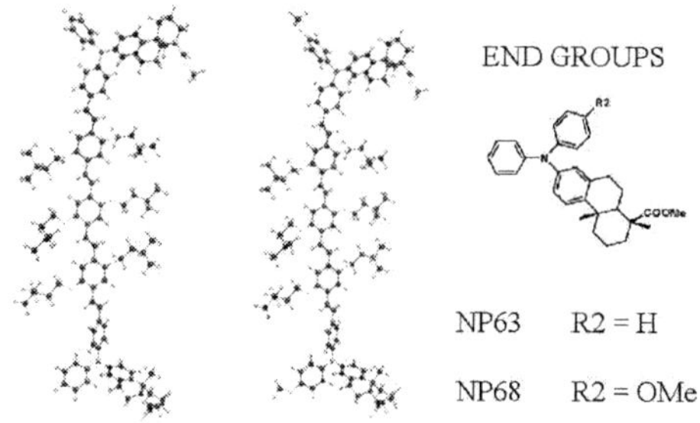

Figure 13. NP63 and NP68 structures.

Since molecular systems with side chains in only one side of the rings should suffer, by the same token, from the same problem, we used the previously computed "minimal" viscosity to estimate the relaxation time of 3,4',4''-

Table 2. Experimental and Calculated (from equation (8)) Values of Relaxation Times of EHOPV3 in Pure non Polar Solvents

trimer	solvent	$\tau(\exp)^{a}$ (ps)	$\tau^{b}_{\mathrm{model}}(C_6)$ (ps)	$\tau_{\mathrm{model}}(C_3)$ (ps)
EHOPV3	pentane	17	16	17
EHOPV3	hexane	28	20	28
EHOPV3	nonane	34	35	34
EHOPV3	decane	41	41	41
EHOPV3	undecane	48	48	48
EHOPV3	dodecane	58	55	58
EHOPV3	tetradecane	69	69	69
EHOPV3	toluene	31	36	31
EHOPV3	MCH	37	39	37

[a] from reference [10] [b] from reference [16]

Table 3. Experimental and Calculated (from equation (8)) Values of Relaxation Times of EHOPV3 in Pure non Polar Solvents

trimer	solvent	$\tau(\exp)^{a}$ (ps)	$\tau^{b}_{\mathrm{model}}(C_6)$ (ps)	$\tau_{\mathrm{model}}(C_3)$ (ps)
NP68	toluene	47	49	45
NP68	decane	54	54	55
NP68	dodecane	66	69	66
NP68	tetradecane	87	83	86
NP63	toluene	40	40	40

[a] from reference [15] [b] from reference [16]

trihexyl-5,5”-dihydro-2,2’:5’,2”-terthiophene, figure 16. The obtained values suggest 1 ps or less if the torsional relaxation PES has similar stiffness to the PPV. A two fold softer PES will produce a relaxation time of around 2.5 picoseconds and an extreme 4 fold softer normal mode will still have a relaxation time below the 5 ps mark. Experimental studies, being conducted by Maçanita

Table 4. Optimized parameters from fitting the model to relaxation times of PV trimers

Trimer	$\eta_{particle}$ (mPa.s)	K' (MPa)
MBOPV3	2.60	48.9
NP63	2.60	22.2
NP68	2.60	19.5
EHOPV3	3.12	29.3

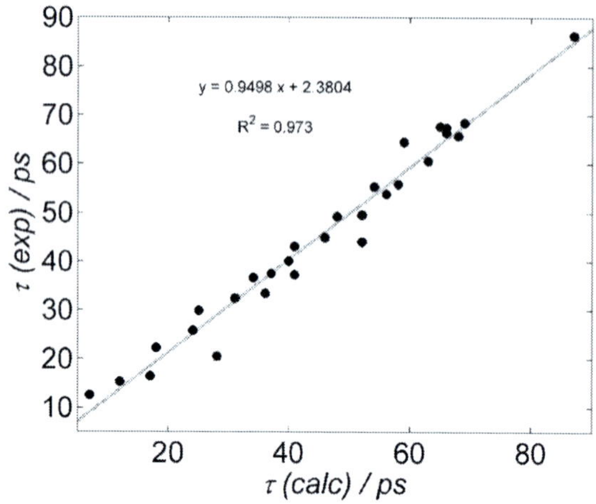

Figure 14. Relaxation times correlation between experimental and model (particle surface potential).

and Seixas de Melo,[4] suggest that a sub-ps value (on the order of a few hundred fs), lower than the former hypothesis, is the correct one.

7. Conclusion

A model, developed for interpreting torsional relaxation in conjugated polymers, has recently been proposed by the authors [16], but it relied heavily on the molecular structure of both the solvent and the solute. Application of the earlier

[4]private communication of preliminary results

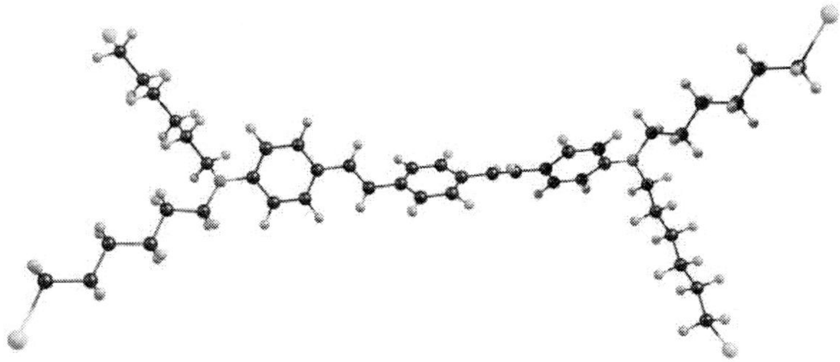

Figure 15. DSBI molecular structure.

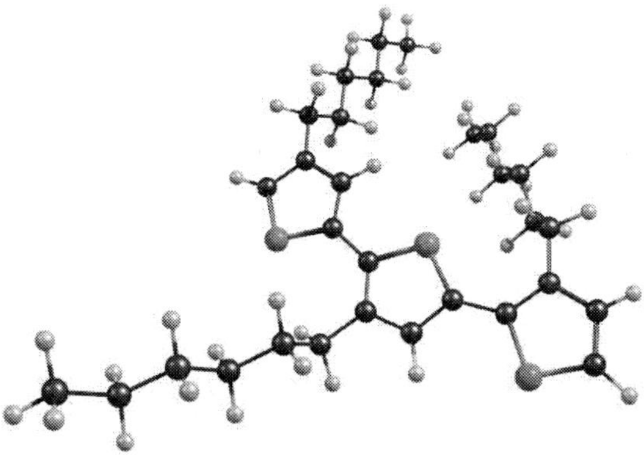

Figure 16. 3,4',4"-trihexyl-5,5"-dihydro-2,2':5',2"-terthiophene.

version required optimization (through the least-squares criterion) of three main parameters: The entropic viscosity of the solute, the stress induced by photo-excitation and a cohesion parameter to account for weaker interactions between different molecules such as solute and solvent. In this work we have sought to remove the necessity of the cohesion parameter as well as any structural information for the solute except for its dielectric constant at high and low frequency (the later just for polar solvents). The data produced by the new model were compared with experimental values as well as previously published material. The new model with just two optimized parameters succeeded not only in reproducing well the relaxation times of PPV trimers but also in predicting viscosities of both polar and non polar solvents, making the model applicable to any inert solvent (one that does not interact chemically with the solute). The influence of lateral chains in the relaxing groups was extended to cover single sided chains or no chains at all, overcoming the earlier restriction to a sticky boundary condition. Introducing slip by allowing the side chains "dash pot" in the mechanical model to drop its viscosity to a value near zero enabled us to understand the microscopic nature of relaxation in the sub-ps time frame.

Acknowledgment

The authors are grateful to the "Fundação para a Ciência e Tecnologia" (Project PEst-OE/QUI/UI0100/2013) and Research Council of Norway (Project No. 191102) for providing financial aid in support of this work.

8. Appendix

8.1. Form Factors

A description of the rational behind the form factors F is fully describbed in [16]. F is basically made proportional to the transverse component of the polarizability tensor depending only on the depolarization ratio L $\big(F = 2/(1-L)\big)$. To compute L it is necessary to assume an approximate shape for our molecules, and a prolate ellipsoid fits a large class of molecules. The minor axis of revolution is assigned to the geometric mean of the two smaller axes (computable from the inertia moments). For this particular shape L is a function of the incomplete elliptic integral of the first kind which, for a prolate ellipsoid, is:

$$I = \int_0^\infty \frac{dx}{(\beta^2+1)\sqrt{1+x}} = \frac{1}{\sqrt{1-\beta^2}} \ln \frac{1+\sqrt{1-\beta^2}}{1-\sqrt{1-\beta^2}},$$

where β is the ratio between the minor and major axes of revolution.

$$L = \left(1 - \frac{1}{1-\beta^2}\right)\left(1 - \frac{I}{2}\right)$$

8.2. Stochastic Differential Equations

The equation which forms the core of our analysis, namely

$$\beta\frac{d\chi}{dt} = -\,\omega^2\chi + A(t) \tag{25}$$

is formally identical with what is known as the Langevin equation, which can be written as

$$m\frac{dv}{dt} = -\zeta v + X(t), \tag{26}$$

where m is the mass of a massive Brownian particle subject to random elastic collisions with a one-dimensional bath (or medium) of very much lighter host molecules. The collective influence of collisions by the medium is represented by the random force X, ζ is the friction coefficient, and v denotes the one-dimensional velocity of the Brownian particle. The purpose of this appendix is to show how equation (26) was handled by Uhlenbeck and Ornstein (hereafter U&O), and to present two simpler methods of obtaining their results.

Upon dividing equation (26) by m, and setting

$$\beta \equiv \frac{\zeta}{m} \quad \text{and} \quad \mathcal{A} \equiv \frac{X}{m}. \tag{27}$$

one gets the neater form

$$\frac{dv}{dt} = -\beta v + \mathcal{A}(t). \tag{28}$$

Let us pause a little to consider the corresponding deterministic equation, satisfied by a variable y:

$$\frac{dy}{dt} = -ky + F(t), \tag{29}$$

where k is a positive constant and F is a *known* function of time. If the initial value of y is y_0, the solution to equation (29) can be expressed as

$$y = y_0 e^{-kt} + e^{-kt}\int_0^t e^{kt_1}F(t_1)dt_1. \tag{30}$$

We note that if F has a constant value, say C, the solution takes the form shown below:

$$y = y_\infty + (y_0 - y_\infty)e^{-kt}, \qquad (y_\infty \equiv C/k). \tag{31}$$

To get over the preliminaries, we note also that, given a definite integral

$$I(t) = \int_0^t e^{\beta t_1} \mathcal{A}(t_1) dt_1, \tag{32}$$

one can write

$$[I(t)]^2 = \int_0^t \int_0^t e^{\beta(t_1+t_2)} \mathcal{A}(t_1)\mathcal{A}(t_2) dt_1 dt_2, \tag{33}$$

We return to equation (28) and follow U&O by integrating it formally, which yields

$$v = v_0 e^{-\beta t} + e^{-\beta t} \int_0^t e^{\beta t_1} \mathcal{A}(t_1) dt_1, \qquad [v_0 \equiv v(0)]. \tag{34}$$

The fact that $\mathcal{A}(t_1)$ on the right-hand side of equation (34) is a random function did not worry U&O unduly, since they were going to attribute physical significance only to averaged quantities. U&O were interested in finding all the moments of v, $\overline{v^n}^{v_0}$ $(n = 1,2,\ldots)$; $\overline{v^n}^{v_0}$ denotes the ensemble average of v^n for particles which start with the same initial velocity v_0.

If one takes an ensemble average of equation (34), one finds

$$\overline{v}^{v_0} = v_0 e^{-\beta t} + e^{-\beta t} \int_0^t e^{\beta t_1} \overline{\mathcal{A}(t_1)}^{v_0} dt_1. \tag{35}$$

In order to proceed further, one needs the value of $\overline{\mathcal{A}(t_1)}^{v_0}$. U&O set

$$\overline{\mathcal{A}(t)}^{v_0} = 0, \tag{36}$$

and obtained the result

$$\overline{v}^{v_0} = v_0 e^{-\beta t}. \tag{37}$$

To obtain $\overline{v^2}^{v_0}$ $(n = 1,2,\ldots)$, U&O squared both sides of equation (35) and calculated the ensemble average of the resulting equation, obtaining

$$\overline{v^2}^{v_0} = v_0^2 e^{-2\beta t} + \int_0^t \int_0^t e^{\beta(t_1+t_2)} \overline{\mathcal{A}(t_1)\mathcal{A}(t_2)}^{v_0} dt_1 dt_2 \tag{38}$$

As before, no progress can be made without specifying the value of $\overline{\mathcal{A}(t_1)\mathcal{A}(t_2)}^{v_0}$. As explained elsewhere [46], U&O did not use Dirac's delta calculus explicitly, and proceeded in a slightly different manner from that used here. We will state their choice by availing ourselves of the delta function:

$$\overline{\mathcal{A}(t_1)\mathcal{A}(t_2)}^{v_0} = a_1 \delta(t_2 - t_1), \tag{39}$$

where a_1 is a constant whose value cannot be determined without an extra input (see below). With this choice, the double integral on the right-hand side of equation (38) can be integrated without any fuss, and some obvious rearrangement immediately gives

$$\overline{v^2}^{v_0} = \frac{a_1}{2\beta} + \left(v_0^2 - \frac{a_1}{2\beta}\right) e^{-2\beta t}. \tag{40}$$

Let us note that, in order to reach the last result, we have made two assumptions about the statistical properties of the random acceleration, which are embodied in equation (36) and (39). Returning to the random force, we may state these properties in the following words: the random force is of mean zero and its spectrum is white.

To find the value of the constant a_1, U&O appealed to the equipartition theorem and claimed that as $t \to \infty$, $\overline{v^2}^{v_0} \to \Theta_B$, where Θ_B is a convenient abbreviation for $k_B T/m$; this led them to the result stated below:

$$a_1 = 2\beta\Theta_B. \tag{41}$$

At this point, it is convenient to introduce a new variable,

$$V = v - \overline{v}^{v_0}, \tag{42}$$

and focus on its moments, called central moments.

To find the higher central moments, U&O proceeded in an analogous manner, and encountered triple, quadruple and higher multiple integrals, which cannot be evaluated without making at least one more assumption, and they assumed that the random force is gaussian, which means that all odd moments of V vanish and all even moments higher than the second can be related to the second central moment:

$$\overline{V^{2n}}^{v_0} = 1 \cdot 3 \cdot 5 \cdots \left[\overline{V^2}^{v_0}\right]^n. \tag{43}$$

8.3. A Simpler Alternative

We will now show that the above results can be obtained more simply. Instead of solving the Langevin equation, and averaging v^n, it will be more profitable to multiply the equation by nv^{n-1}, replace $nv^{n-1}dv/dt$ by dv^n/dt, and average the resulting equation. For $n = 1$, we arrive, after setting $\overline{\mathcal{A}}^{v_0} = 0$, at the equation

$$\frac{d\overline{v}^{v_0}}{dt} = -\beta\overline{v}^{v_0}, \tag{44}$$

whose solution is

$$\overline{v}^{v_0} = v_0 e^{-\beta t}. \tag{45}$$

This is, of course, the same result as that stated in equation (37), obtained by taking a v_0-average of equation (34) and using equation (37).

Averaging the equation found by multiplying equation (28) by $2v$ gives

$$\frac{d\overline{v^2}^{v_0}}{dt} = -2\beta\overline{v^2}^{v_0} + 2\overline{v\mathcal{A}}^{v_0}, \tag{46}$$

which is an equation of the same form as equation (29). Since we know that $\overline{v^2}^{v_0} \to \Theta_\mathrm{B}$ at long times, we must assume that the second term on the right-hand side is independent of time, and using the solution stated in equation (31), we are led to the result

$$\overline{v\mathcal{A}}^{v_0} = \beta\Theta_\mathrm{B} = \frac{a_1}{2}, \tag{47}$$

which yields the desired time-dependence of $\overline{v^2}^{v_0}$:

$$\overline{v^2}^{v_0} = \Theta_\mathrm{B} + \left(v_0^2 - \Theta_\mathrm{B}\right) e^{-2\beta t}. \tag{48}$$

Having found $\overline{v}^{v_0}$ and $\overline{v^2}^{v_0}$, we can calculate the variance $\sigma^2 = \overline{v^2}^{v_0} - (\overline{v}^{v_0})^2$, which comes out to be

$$\sigma^2 = \Theta_B \left(1 - e^{-2\beta t}\right), \tag{49}$$

and if we assume that the governing probability density function is a gaussian, its explicit form can be immediately stated:

$$G(v,t;v_0) = \left[\frac{1}{2\pi\Theta_\mathrm{B}\left(1-e^{-2\beta t}\right)}\right]^{1/2} \exp\left\{-\frac{1}{2}\frac{\left(v - v_0 e^{-\beta t}\right)^2}{\Theta_\mathrm{B}(1-e^{-2\beta t})}\right\}. \tag{50}$$

8.4. Conversion to a Difference Equation

U&O mentioned that the time dependent averages can be calculated from the results obtained by de Haas-Lorentz [47]. She integrated equation (28) from $t = (N-1)\tau$ to $t = N\tau$, thereby converting it into a difference equation:

$$v_N - v_{N-1} = -\beta v_{N-1}\tau + I_{N-1}, \tag{51}$$

where

$$I_{N-1} = \int_{(N-1)\tau}^{N\tau} \mathcal{A}(t_1)dt_1. \tag{52}$$

Equation(51) describes a random walk in v-space at discrete instants of time, and its solution requires only summation of geometric series, as demonstrated first by de Haas-Lorentz [47]. The time interval τ is so short that v does not change substantially during this interval, but the fluctuations in $\mathcal{A}$ are supposed to be extremely rapid even on this short time scale. With this background, we can already state two statistical properties of the random term on the right-hand side of equation (51):

$$\overline{I_k}^{v_0} = 0 \tag{53}$$

$$\overline{I_j I_k}^{v_0} = \delta_{ij}\Pi_1, \quad \delta_{ij} = \begin{cases} 1, & \text{if } j = k, \\ 0, & \text{otherwise.} \end{cases} \tag{54}$$

It follows from iteration of equation (51) that v_N can be arranged as a sum. Since this section is supposed to provide a discrete variant of the approach used by U&O, we will express the sum so as to highlight the correspondence between equation (34) and its discrete counterpart shown below:

$$v_N = v_0\rho^N + \rho^N \sum_{k=1}^{N} I_{k-1}\rho^{-k}, \tag{55}$$

where

$$\rho \equiv (1 - \beta\tau). \tag{56}$$

Let us now take an average of equation (55), and arrive, after using equation (53), at the relation

$$\overline{v_N}^{v_0} = v_0\rho^N \tag{57}$$

We would expect, on letting $N \to \infty$ and $\tau \to 0$, to recover the results found by U&O if we make the identifications $N\tau = t$, so that $v_N = v(t)$, $x_N = x(t)$, and so on. Since we are going to let $\tau \to 0$, it is permissible to write

$$\rho = e^{-\beta\tau}, \quad \text{or} \quad \rho^N = e^{-\beta t}. \tag{58}$$

We have now established that, as $\tau \to 0$, one gets

$$\overline{v}^{v_0} = v_0 e^{-\beta t}, \tag{59}$$

a result that coincides with that stated in equation (37).

Upon squaring-and-averaging, equation (55), and making use of equation (53) and equation (54), we are left with

$$\begin{aligned}\overline{v_N^2}^{v_0} &= v_0^2\rho^{2N} + \rho^{2N}\Pi_1 \sum_{k=0}^{N-1} \rho^{-2k} \\ &= v_0^2\rho^{2N} + \Pi_1 \left[\frac{\rho^{2N}-1}{1-\rho^{-2}}\right]. \end{aligned} \tag{60}$$

As we let $\tau \to 0$, the first term on the right-hand side goes into $v_0^2 e^{-2\beta t}$, whereas the factor within the square brackets will approach

$$\frac{1}{2\beta\tau}(1-e^{-2\beta t}).$$

Since $\overline{v^2}^{v_0}$ must tend to Θ_B as $t \to \infty$, we choose the limiting value

$$\lim_{\tau\to 0} \frac{\Pi_1}{\tau} = 2\beta\Theta_B, \tag{61}$$

and arrive thereby at the result appearing in equation (48).

8.5. Constitutive Relations for Viscoelastic Materials

To obtain constitutive relationships between stresses imposed on materials and the corresponding strains we need to solve equations (62) and (63)

$$\sigma(t) = \int_{-\infty}^{\infty} r(t-\tau)\, \frac{d\chi(\tau)}{dt}\, d\tau, \tag{62}$$

$$\chi(t) = \int_{-\infty}^{\infty} c(t-\tau)\, \frac{d\sigma(\tau)}{dt}\, d\tau, \tag{63}$$

where $c(t-\tau)$ and $r(t-\tau)$ are the creep and relaxation functions respectively.

Before going on on further details we must discuss the integration limits on equations (62) and (63). In a *causal* system the response can only be a function of present and past events and never depends on future events which means the systems are not anticipative. In that way the upper limit of both integrals should be changed to the current time t. The other problem is *memory*; some systems depend only on the values of system variables at present time while other systems can remember events from the near past and, sometimes, keep record of everything from all its existence.

Lets consider the following kernel for the relaxation function:

$$r(t) = Gh(t) + \eta\delta(t) \tag{64}$$

where $h(t)$ and $\delta(t)$ are the unit-step and unit-impulse functions, respectively. Introducing it in equation (62) we get equation (65)

$$\sigma = G\chi + \eta\dot{\chi} \tag{65}$$

We can now easily recognize in the first term of the right side of equation (65) the expression for the harmonic force of a spring and in the second term the viscous force exerted by a fluid on a moving object. Elasticity and viscosity are by virtue of equation (64) in a single foot: Viscosity is the response of a memoryless system (the unit-impulse is a single time selection function) and elasticity is an infinite memory (elephant) one (the unit-step is an integrator).

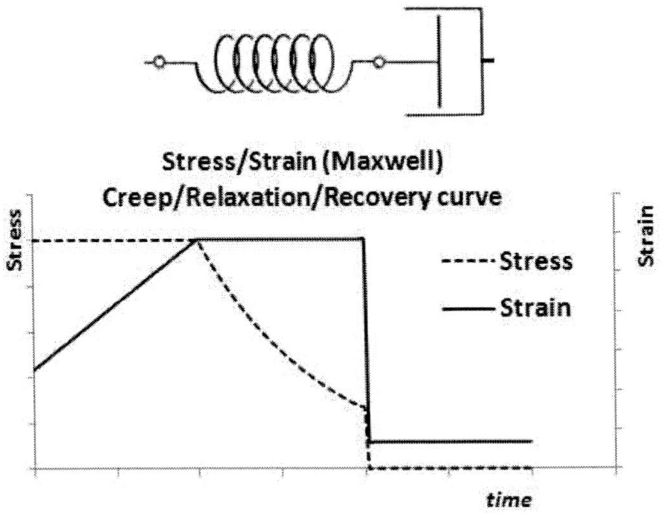

Figure 17. Maxwell model simulation of creep/relaxation/recovery.

Since the two right terms of equation (64) are beacons to the two extreme systems, we will try to describe a general viscoelastic behavior by a linear combination of the two extremes representing the infinite memory by a spring symbol and the memoryless by a dash pot.

Lets now consider the following essay:

1 - At time t_o apply a constant stress σ_o (creep)
2 - At time t_1 let the system relax at constant strain (Relaxation)
3 - At time t_2 remove the stress and allow the system to recover (Recovery)

The results of the essay for series and parallel combinations of springs and dash pots will be represented in figures 17 and 18.

Maxwell Model

The series association of theses two elements creates a Maxwell model. The applied stress is felt equally by both elements but each one has its own share of the deformation. After the application of the stress ($t = 0$) the spring absorbs all the strain, but then ($t > 0$) there is an irreversible build up of strain in the dash pot. At time t_1 the strain is held constant and the stress relaxes. If at a time t_2 we remove the stress only the strain that remains in the spring can be recovered.

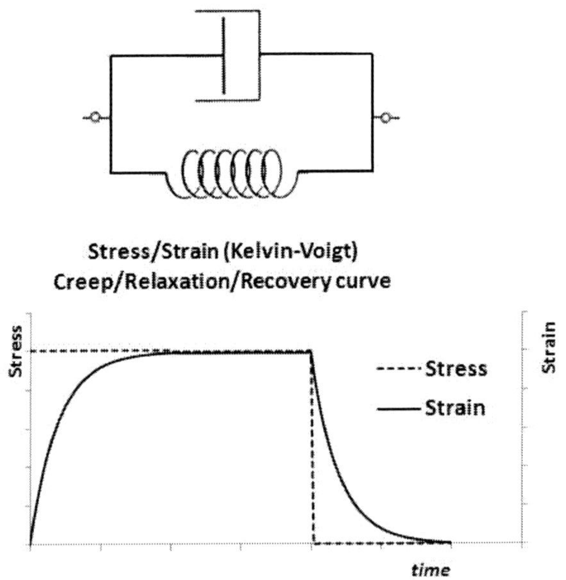

Figure 18. Kelvin-Voigt model simulation of creep/relaxation/recovery.

Kelvin-Voigt Model

The parallel association of a spring and a dash pot creates a Kelvin-Voigt model. The strain is the same in both elements but each one feels a different stress. After the application of the stress ($t = 0$) the dash pot takes all the stress, but as strain builds up ($t > 0$) there is a transfer of stress to the spring. This means there is a retardation time between the application of a stress and the development of a strain. If the stress is removed the dash pot will held the strain but with time the stress stored in the spring will bring the system to the original unstrained configuration.

References

[1] Hadziioannou, G.; Malliaras, G. *Semiconducting polymers: chemistry, physics and engineering*; Semiconducting Polymers: Chemistry, Physics and Engineering v. 2; Wiley-VCH, 2007.

[2] Kamat, P. V. *J. Phys. Chem. C* 2007, 111, 2834–2860.

[3] Borchardt, J. K. *Mater. Today* 2004, 7, 42 – 46.

[4] Wells, N. P.; Boudouris, B. W.; Hillmyer, M. A.; Blank, D. A. *J. Phys. Chem. C* 2007, 111, 15404–15414.

[5] Rodrigues, R. F.; Charas, A.; Morgado, J.; Maçanita, A. *Macromolecules* 2010, 43, 765–771.

[6] Ferreira, B.; da Silva, P. F.; Sergio Seixas de Melo, J.; Pina, J.; Maçanita, A. *J. Phys. Chem. B* 2012, 116, 2347–2355.

[7] Dias, F.; Maçanita, A.; de Melo, J.; Burrows, H.; Guntner, R.; Scherf, U.; Monkman, A. *J. Chem. Phys.* 2003, 118, 7119–7126.

[8] Dykstra, T. E.; Hennebicq, E.; Beljonne, D.; Gierschner, J.; Claudio, G.; Bittner, E. R.; Knoester, J.; Scholes, G. D. *J. Phys. Chem. B* 2009, 113, 656–667.

[9] Clark, J.; Nelson, T.; Tretiak, S.; Cirmi, G.; Lanzani, G. *Nature Phys.* 2012, 8, 225–231.

[10] Di Paolo, R. E.; de Melo, J. S.; Pina, J.; Burrows, H. D.; Morgado, J.; Maçanita, A. L. *ChemPhysChem* 2007, 8, 2657–2664.

[11] Sluch, M.; Godt, A.; Bunz, U.; Berg, M. *J. Am. Chem. Soc.* 2001, 123, 6447–6448.

[12] Dias, F. B.; Pollock, S.; Hedley, G.; Palsson, L.-O.; Monkman, A.; Perepichka, I. I.; Perepichka, I. F.; Tavasli, M.; Bryce, M. R. *J. Phys. Chem. B* 2006, 110, 19329–19339.

[13] Hintschich, S. I.; Dias, F. B.; Monkman, A. P. *Phys. Rev. B* 2006, 74, 45210–45219.

[14] Di Paolo, R. E.; Burrows, H. D.; Morgado, J.; Maçanita, A. L. *ChemPhysChem* 2009, 10, 448–454.

[15] Di Paolo, R. E.; Gigante, B.; Esteves, M. A.; Pires, N.; Santos, C.; Lameiro, M. H.; Seixas de Melo, J.; Burrows, H. D.; Maçanita, A. L. *ChemPhysChem* 2008, 9, 2214–2220.

[16] Galvão, A. M.; Di Paolo, R. E.; Maçanita, A. L.; Naqvi, K. R. *ChemPhysChem* 2013, 14, 583–590.

[17] Crawford, R. *Plastics Engineering*; Elsevier Science, 1998.

[18] Felderhof, B. U. *J. Chem. Phys.* 2011, 135, 234901(1)–234901(6).

[19] Magee, M. D. *J. Chem. Soc. Farad. T 2* 1974, 70, 929–938.

[20] Dyre, J. C.; Wang, W. H. *J. Chem. Phys.* 2012, 136, 224108(1)–224108(5).

[21] Kubo, R. *Rep. Prog. Phys.* 1966, 29, 255–284.

[22] Uhlenbeck, G.; Ornstein, L. *Phys. Rev.* 1930, 36, 0823–0841.

[23] Ornstein, L.; Wijk, W. V. *Physica* 1934, 1, 235 – 254.

[24] Ornstein, L.; van Wijk, W. *Physica* 1934, 1, 966 – 966.

[25] SanMiguel, M.; Sancho, J. *J. Stat. Phys.* 1980, 22, 605–624.

[26] Smoluchowski, M. V. *Phys. Z.* 1916, 17, 557–571.

[27] Dyre, J.; Olsen, N.; Christensen, T. *Phys. Rev. B* 1996, 53, 2171–2174.

[28] Ewell, R. H.; Eyring, H. *J. Chem. Phys.* 1937, 5, 726–736.

[29] Hao, T. *Adv. Colloid Interfac.* 2008, 142, 1–19.

[30] Zwanzig, R.; Mountain, R. D. *J. Chem. Phys.* 1965, 43, 4464–4471.

[31] Tang, K. *Phys. Rev.* 1969, 177, 108–114.

[32] Wohlfarth, C.; Wohlfarth, B. In *Viscosity of Pure Organic Liquids and Binary Liquid Mixtures: pure organic liquids*; Lechner, M. D., Ed.; IV/18B; Springer, 2001.

[33] Iwahashi, M.; Kasahara, Y. *J. Oleo Sci.* 2007, 56, 443–448.

[34] Oliveira, M. J. T.; Botti, S.; Marques, M. A. L. *Phys. Chem. Chem. Phys.* 2011, 13, 15055–15061.

[35] Hamaker, H. *Physica* 1937, 4, 1058 – 1072.

[36] Chew, W. C.; Sen, P. N. *J. Chem. Phys.* 1982, 77, 4683–4693.

[37] Chew, W. C.; Sen, P. N. *J. Chem. Phys.* 1982, 77, 2042–2044.

[38] Rouse, P. *J. Chem. Phys.* 1953, 21, 1272–1280.

[39] Padding, J.; Briels, W. *J. Chem. Phys.* 2001, 114, 8685–8693.

[40] Doi, M.; Edwards, S. F. *The Theory of Polymer Dynamics*; Clarendon, Oxford, 1986.

[41] Doi, M. *J. Polym. Sci. Pol. Lett.* 1981, 19, 265–273.

[42] Doi, M. *J. Polym. Sci. Pol. Phys.* 1983, 21, 667–684.

[43] Schmidt, M. W.; Baldridge, K. K.; Boatz, J. A.; Elbert, S. T.; Gordon, M. S.; Jensen, J. H.; Koseki, S.; Matsunaga, N.; Nguyen, K. A.; Su, S.; Windus, T. L.; Dupuis, M.; Montgomery, J. A. *J. Comput. Chem.* 1993, 14, 1347–1363.

[44] Gordon, M. S.; Schmidt, M. W. In *Theory and Applications of Computational Chemistry: the first forty years*; Dykstra, C. E., Frenking, G., Kim, K. S., Scuseria, G. E., Eds.; Elsevier: Amsterdam, 2005; pp 1167–1189.

[45] Naqvi, K. R. *J. Chem. Educ.* 1989, 66, 703.

[46] Naqvi, K. R. (2005). The origin of the Langevin equation and the calculation of the mean squared displacement: Let's set the record straight. arXiv:physics/0502141v1 [physics.chem-ph].

[47] de Haas-Lorentz, G. *Die Brownsche Bewegung und einige verwandte Erscheinungen*; Die Wissenschaft; 1913.

In: Photochemistry: New Research
Editors: A. B. Smith et al, pp. 77-111
ISBN 978-1-62808-615-7

Chapter 4

QUENCHING OF FLAVIN TRIPLETS BY AMINO ACIDS AND HYDROPHILIC CAROTENOIDS: INVESTIGATIONS BASED ON LASER FLASH PHOTOLYSIS AND E-TYPE DELAYED FLUORESCENCE

K. Razi Naqvi**, *Heng Li, Øyvind Persvik and T. B. Melø
Department of Physics, Norwegian University
of Science and Technology (NTNU), Trondheim, Norway

Abstract

The photochemical behavior of the flavin chromophore is a matter of considerable interest, not only because the chromophore is an integral part of numerous flavoproteins, but also because riboflavin, or vitamin B_2, is an essential micronutrient and a component of light-sensitive beverages such as milk and beer. The photodegradation of riboflavin-containing beverages is attributed to the generation (following absorption of light and intersystem crossing) of triplet flavin, which sensitizes the formation of singlet oxygen, a highly toxic species. Since the flavin triplet is known to be quenched by suitable substrates (for example the amino acids tyrosine and tryptophan), sensitization of singlet oxygen can be blocked by the addition of an appropriate substrate. According to previous reports, tryptophan and tyrosine quench triplet flavin, but crocin (a naturally occurring

*E-mail address: razi.naqvi@ntnu.no

hydrophilic carotenoid) does not; the quenching occurs through electron transfer from the quencher to the flavin triplet and the rate constant of quenching in aqueous buffer is rather large ($> 10^9$ M^{-1} s^{-1}). The present study was motivated by two considerations: Firstly, since carotenoids are known to be good electron donors and efficient triplet quenchers, the reported failure of crocin to quench the flavin triplet warrants a reinvestigation of the quenching kinetics; secondly, studies conducted over the last four decades have led to widely discordant values for the rate constants of the processes which govern the decay of triplet flavin in aqueous media, and there is clearly a need to identify the source of the discrepancies. The present study has revealed that crocin is nearly as good a quencher as tryptophan, and that quenching of flavin triplets by crocin and two other hydrophilic carotenoids proceeds through two parallel channels: electron transfer from the carotenoid to triplet flavin and energy transfer from triplet flavin to the carotenoid. For tracking the concentration of triplet flavin, detection of delayed fluorescence provides a far more convenient alternative (to laser flash photolysis) in general, and becomes particularly advantageous in systems containing carotenoids, where absorption by the carotenoid radical cation (produced through direct photoionization as well as electron transfer) overlaps that by triplet flavin. The practical advantages of using naturally occurring hydrophilic carotenoids (such as crocin and norbixin) for the photoprotection of riboflavin-containing beverages are also discussed.

PACS 05.45-a, 52.35.Mw, 96.50.Fm

Keywords: Riboflavin, lumiflavin, triplet, quenching, delayed fluorescence

1. Introduction

Flavins—substances containing the isoalloxazine ring—are widespread in biological systems and act as oxidases [1], electron carriers and blue light receptors [2]. Three of the best known members of the flavin family are lumiflavin, riboflavin and FMN, which differ from one another in having a different side group (R) attached to N10 of the isolloxazine ring (see Figure 1). In this chapter, the name vitamin B_2, which is sometimes used as a synonym for riboflavin and sometimes as a generic label for all biologically active flavins, will be used only in the latter sense. The principal members of the vitamin B_2 group found in nature are riboflavin, riboflavin-5′-phosphate (more commonly called flavin mononucleotide and abbreviated as FMN) and riboflavin-5′-adenosylphosphate (better known under the inappropriate name flavin adenine dinucleotide and abbreviated as FAD). Riboflavin is a constituent of the electron transport chain in the cellular respiratory system, with the consequence that organisms exposed to sunlight are subjected to the photochemistry of the flavins. For further information concerning the biopotency and physicochemical properties of these three molecules, the reader is referred to Chapter 8 of Ball's monograph [3].

The early interest in flavin photochemistry was triggered by the involvement of photo-initiated reactions in phototaxis and bioluminescence [4]. FMN or FAD is the chromophore in three out of the altogether six groups of photoreceptor proteins, which are: the cryptochromes, the phototropins and the family BLUF (sensors for blue light using FAD) [5]. The initial event in flavin-containing photoreceptor proteins is light absorption by the non-covalently bound flavin chromophore, and subsequent photochemical events change the apoprotein into a signalling state. Common to the cryptochrome reactions [6], which might range from synchronization of the circadian clock in animals to seed germination in plants, is the formation of the FAD triplet state through intersystem crossing, and the production, as a result of the transfer of an electron from a neighbouring tryptophan, of a semireduced FAD, which induces the signalling state. A further reduction produces the fully reduced FAD, which turns off signalling. In the phototropin family [7], light absorption in plants mediates responses like chloroplast movement and stomatal opening. In this case a hydrogen atom is transferred from a nearby cysteine to triplet flavin, forming a covalent adduct between the C4a atom in the flavin and the sulphur atom of cysteine in the protein. This binding changes the planar configuration of the isoalloxazine nucleus, and this change is transmitted to the LOV (these proteins are regulated by light, oxygen or voltage) domain of the protein, which is effecting the signal. The BLUF family [8] is predominantly found in prokaryotes, where it regulates expression of photosynthesis genes. Light absorption induces an electron transfer from a conserved tyrosine residue to the flavin, which causes the side chain of a nearby glutamine to rotate, followed by electron back-transfer to tyrosine. The tyrosine rotamer induces new hydrogen bonds, a change that causes the protein to enter a signalling state.

In bacterial luminescence, which is catalyzed by luciferase, two simple substrates—a reduced flavin mononucleotide ($FMNH_2$) and a long-chain aliphatic aldehyde (RCHO)—are oxidized by molecular oxygen and luciferase, a reaction that liberates excess energy in the form of blue-green light (490 nm). The glow is persistent because although the aldehyde is consumed in the reaction, it is continuously synthesized by the bacterium.

A useful review of the photochemical and photophysical properties of flavins was published by Heelis in 1982 [4]. For a more recent survey of the photochemical and photobiological properties of flavins, the reader is referred to a comparatively recent book edited by Hader and Jori [9].

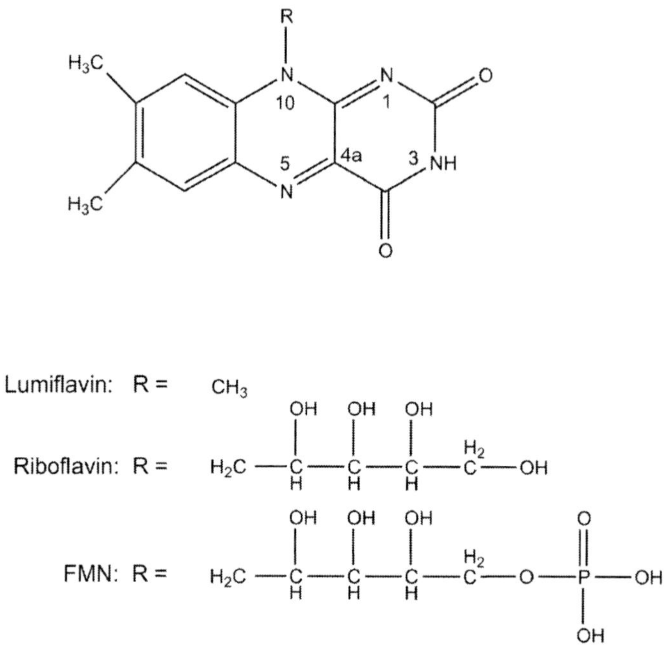

Figure 1. Molecular structures of three common flavins.

2. Notation, Background and Motivation for the Study

We will use the symbol **F** to denote a flavin molecule, and add the superscripts zero, dagger and asterisk (0, † and $*$) when it is our intention to specify that the molecule occupies the lowest singlet state (S_0), the lowest triplet state (T_1) and the first singlet excited state (S_1), respectively. A dioxygen molecule will be denoted by the standard symbol (O_2), when it is in its ground state (a triplet) and by O_2^* when it is in the first singlet excited state. When it becomes necessary to distinguish between the three flavins shown in Figure 1, we will add a pre-superscript; thus $^{L}\mathbf{F}$, $^{R}\mathbf{F}$, $^{M}\mathbf{F}$ will denote lumiflavin, riboflavin and FMN, respectively. The symbols **C**, **S** and Q will be used for denoting a carotenoid, a substrate and a (deliberately added) quencher, respectively; thus Q may stand also for a carotenoid or a substrate.

Photoexcitation of a flavin molecule raises it to a vibronic state in the singlet manifold, from where it falls to the S_1 state in a time that is so short on our time scale that the process may be called instantaneous. Thus it will be sufficient for us to assume that photoexcitation always promotes the molecule to the S_1 state, and it may therefore be represented as process 1 in Scheme 1, which also lists the subsequent unimolecular processes of interest in the present context.

SCHEME 1. UNIMOLECULAR PROCESSES OF INTEREST

Photoexcitation:	$h\nu + \mathbf{F}^0 \longrightarrow \mathbf{F}^*$	(1)
Fluorescence:	$\mathbf{F}^* \xrightarrow{k_{FL}} \mathbf{F}^0 + h\nu_F$	(2)
Internal Conversion:	$\mathbf{F}^* \xrightarrow{k_{SG}} \mathbf{F}^0$	(3)
Triplet formation (ISC):	$\mathbf{F}^* \xrightarrow{k_{ST}} \mathbf{F}^\dagger$	(4)
Thermal reactivation of S_1:	$\mathbf{F}^\dagger \xrightarrow{k_{TS}} \mathbf{F}^*$	(5)
Nonradiative decay of T_1:	$\mathbf{F}^\dagger \xrightarrow{k_{TG}} \mathbf{F}^0$	(6)
Phosphorescence:	$\mathbf{F}^\dagger \xrightarrow{k_{PH}} \mathbf{F}^0 + h\nu_P$	(7)

In this chapter, the fate of $\mathbf{F}^\dagger$ is our primary concern, and one cannot ignore, in any system of practical interest, bimolecular processes, the most important of which are listed in Scheme 2 (in which TTA is an abbreviation for *triplet-triplet annihilation*).

SCHEME 2. BIMOLECULAR PROCESSES OF INTEREST

$$\text{Quenching by } O_2: \quad O_2 + \mathbf{F}^\dagger \xrightarrow{k_{QX}} O_2^* + \mathbf{F}^0 \tag{8}$$

$$\text{Ion-pair formation via TTA:} \quad \mathbf{F}^\dagger + \mathbf{F}^\dagger \xrightarrow{k_{2a}} \mathbf{F}^{\bullet+} + \mathbf{F}^{\bullet-} \tag{9}$$

$$S_1 \text{ formation via TTA:} \quad \mathbf{F}^\dagger + \mathbf{F}^\dagger \xrightarrow{k_{2b}} \mathbf{F}^* + \mathbf{F}^0 \tag{10}$$

$$\text{Triplet-triplet quenching:} \quad \mathbf{F}^\dagger + \mathbf{F}^\dagger \xrightarrow{k_{2c}} \mathbf{F}^0 + \mathbf{F}^0 \tag{11}$$

$$\text{Ground-state quenching (GSQ):} \quad \mathbf{F}^\dagger + \mathbf{F}^0 \xrightarrow{k_{3a}} \mathbf{F}^0 + \mathbf{F}^0 \tag{12}$$

$$\text{Ion-pair formation via GSQ:} \quad \mathbf{F}^\dagger + \mathbf{F}^0 \xrightarrow{k_{3b}} \mathbf{F}^{\bullet+} + \mathbf{F}^{\bullet-} \tag{13}$$

$$\text{Triplet-triplet transfer:} \quad \mathbf{F}^\dagger + \mathbf{S}^0 \xrightarrow{k_{4a}} \mathbf{F}^0 + \mathbf{S}^\dagger \tag{14}$$

$$\text{Electron transfer:} \quad \mathbf{F}^\dagger + \mathbf{S}^0 \xrightarrow{k_{4b}} \mathbf{F}^{\bullet-} + \mathbf{S}^{\bullet+} \tag{15}$$

$$\text{H-atom abstraction:} \quad \mathbf{F}^\dagger + \mathbf{RH} \xrightarrow{k_5} \mathbf{FH}^\bullet + \mathbf{R}^\bullet \tag{16}$$

If $\mathbf{F}^{\bullet-}$ is generated in an aqueous solution at neutral pH, it rapidly converts to $\mathbf{FH}^\bullet$, the flavin semiquinone radical, but in other solvents $\mathbf{F}^{\bullet-}$ may survive for longer periods. Though the net outcome of process 15 is the formation of $\mathbf{FH}^\bullet$, Tsentalovich *et al.* [10] have shown that the actual mechanism, electron transfer followed by protonation or direct H-atom abstraction, can be established by careful adjustment of experimental conditions (including sufficient time resolution); the proposal previously advanced [11] on the basis of conventional flash photolysis (with a time resolution of a few μs), that tryptophan quenches ${}^{R}\mathbf{F}^\dagger$

through process 15 was incorrect. As for process 14, it is worth noting that, with the exception of carotenoids (see below), the triplet energy level of the substrates heretofore used was higher than that of **F**, a choice that blocked the energy transfer channel $\mathbf{F}^{\dagger} + \mathbf{S}^{0} \rightarrow \mathbf{F}^{0} + \mathbf{S}^{\dagger}$.

A few words will now be expressed to explain the importance of investigating the quenching of flavin triplets in various systems in which the chromophore is not in a protein environment.

Exposure to light degrades milk and other beverages containing riboflavin [12–19]. The proposed remedy involves incorporation of various nontoxic substances, for example amino acids and purine derivatives, capable of quenching $^{R}\mathbf{F}^{\dagger}$. Carotenoids are known to play a dual role in photoprotection: preventing the formation of singlet oxygen by quenching potential sensitizers, and quenching singlet oxygen itself. The vast majority of naturally occurring carotenoids are hydrophobic, which makes them inappropriate for aqueous systems; even so, at least two studies involving the use of hydrophobic carotenoids as quenchers have appeared in the literature [16, 17]. On the other hand, the energy of the lowest triplet state of carotenoids is rather low, which leads one to expect that using a carotenoid (**C**) would allow energy transfer to compete with electron capture by $^{R}\mathbf{F}^{\dagger}$, and that one would observe both energy transfer (process 14) and electron transfer (process 15). However, the only study where the interaction between triplet flavin and the hydrophilic carotenoid crocin (see Figure 2) was investigated resulted in the conclusion that these carotenoids are not able to quench $^{R}\mathbf{F}^{\dagger}$ [16]. The resolution of this contradiction was one of the main motivations for undertaking the study the results of which are described in this chapter.

It will be convenient to follow the notation used in reference [20]; to this end, we introduce some abbreviations:

$$k_{S1} = k_{FL} + k_{SG} + k_{ST} \tag{17}$$

$$\phi^{\dagger} = k_{ST}/k_{S1} \tag{18}$$

$$k_1 = k_{PH} + k_{TG}, \tag{19}$$

$$k_1' = k_1 + k_{QX}[O_2], \tag{20}$$

$$k_2 = k_{2a} + k_{2b} + k_{2c}, \tag{21}$$

$$k_3 = k_{3a} + k_{3b}. \tag{22}$$

We can now write down the rate equation which will form the basis of our

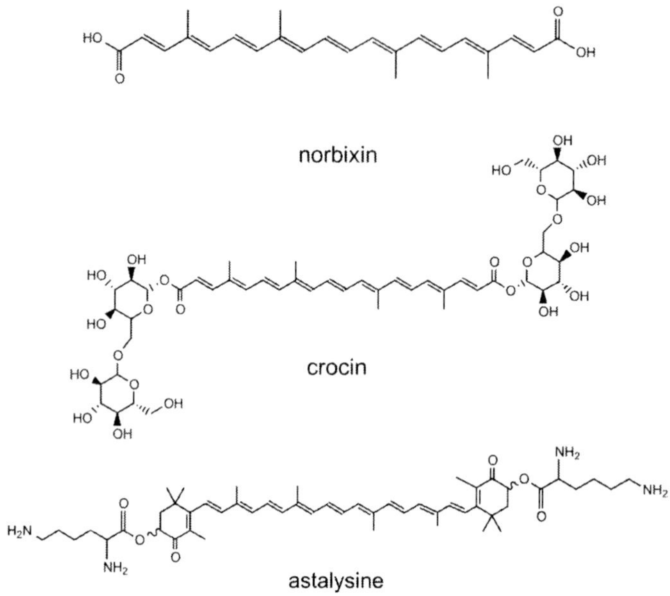

Figure 2. Molecular structures of the hydrophilic carotenoids norbixin, crocin and astalysine.

discussion:

$$\frac{\mathrm{d}[\mathbf{F}^{\dagger}]}{\mathrm{d}t} = -\left(k_1' + k_3[\mathbf{F}^0] + k_q[Q] + k_2[\mathbf{F}^{\dagger}]\right)[\mathbf{F}^{\dagger}]. \tag{23}$$

It is known that, even when all precautions are taken to suppress bimolecular events (by using a high-purity solvent, creating a low triplet population, maintaining a low ground-state concentration, and minimizing the concentration of O_2), the measured triplet lifetime is still determined by deactivation due to adventitious quenchers (mainly O_2). In other words, one does not expect to find good agreement between the values of k_1' determined in different laboratories, and it is worth noting that the symbol k_1 is frequently used, instead of k_1', by authors who are dealing with solutions which have been imperfectly deoxygenated.

In Table 1, we have compiled the values of k_1, k_2, and k_3 obtained by using a conventional flash photolysis apparatus (where path lengths of 10 or 20 cm could be used). The large discrepancies speak volumes for the difficulty of the task and the inadequacy of the tools, and one can only conclude that no confidence can be attached to any of these results. In fact, Grellman and Scholz [21], who used a sampling flash apparatus with a path length of 10 cm showed, as far back as 1979, that extensive signal averaging and digital data analysis are vital for extracting the values of k_1 and k_2 (for anthracene and pyrene in fluid solutions).

Our principal aims are three. The first is to point out that an adequate analysis of the kinetics of flavin triplets in a quencher-free system is not possible in a standard laser photolysis experiment (with a path length of 1 cm). Second, we will show that observation of E-type delayed fluorescence provides a convenient and far superior alternative for obtaining reliable values of the lifetime of triplet flavin (both in the absence and presence of an added quencher); this approach will furnish accurate values k_q for the following quenchers: cysteine, tryptophan, crocin, norbixin and astalysine (see Figure 2). Third, once the values of k_1, and k_3 become available, laser photolysis data can be used for finding k_2, which cannot be determined by the alternative approach.

Table 1. Rate constants determined by using conventional flash photolysis (*l* denotes the optical path length).

Flavin	$\frac{k_1 \text{ or } k_1'}{10^3 \times s^{-1}}$	$\frac{k_2}{10^9 \times M^{-1}\, s^{-1}}$	$\frac{k_3}{10^8 \times M^{-1}\, s^{-1}}$	$\frac{l}{cm}$	Reference
Lumiflavin	0.7	0.89	3.7	10	[20]
Lumiflavin	1.5	1.8	7	10	[22]
Lumiflavin	1.5	1.8	2.4	20	[23]
Lumiflavin	1.5		1.2	10	[24]
Riboflavin	3.2		1.1	10	[24]
FMN	4.9		1.2	10	[24]
Lumiflavin	1.4	4.5	< 0.2	10	[25]

The essence of an optical pump-probe measurement is to produce, by using a light pulse of sufficiently short duration, a large population of an excited state and to measure, after the pulse is over, the change in the absorbance of the sample, $\Delta A(\lambda;t) = \widetilde{A}(\lambda;t) - A(\lambda)$, where $A(\lambda)$ and $\widetilde{A}(\lambda;t)$ denote, respectively, the absorbance of the sample at wavelength λ before excitation and at time t after the pulse. For time scales longer than a few ns, two modes of operation are available: single-wavelength kinetic spectroscopy and time-resolved multichannel spectroscopy. In the former mode, one uses photoelectric detection and a digital oscilloscope to record a trace of ΔA at a pre-selected wavelength λ; in the latter, a multichannel spectrograph is used to record an entire spectrum after a pre-selected value of the delay time time t. In either case, one can write $\Delta(\lambda;t) = \lg[I(\lambda)/\widetilde{I}(\lambda;t)]$, where $I(\lambda)$ and $\widetilde{I}(\lambda;t)$ denote the intensity of the light leaving the sample before and at time t after it is exposed to the pump source.

In what follows, the abbreviation $\Delta@d$ will refer to the spectrum recorded after a delay of $d = t/\mu$s. For a given λ, we will use the abbreviation $\Delta_w(t) = [\lg I_w(t)/\widetilde{I}_w(t)]$ instead of $\Delta A(\lambda;t)$, where $w = \lambda/\text{nm}$. The absorption coefficient of a particular species at wavelength λ will be denoted by $\varepsilon_\lambda^{Ⓢ X}$, with X=L, M or R and Ⓢ $= 0$ for $\mathbf{F}^0$, † for $\mathbf{F}^\dagger$, • for $\mathbf{FH}^\bullet$, and ⊕ for $\mathbf{F}^{\bullet+}$ and $\mathbf{C}^{\bullet+}$; when the numerical value of an absorption coefficient at a given wavelength λ is quoted, the subscript λ will be replaced by w. The abbreviations TmS and DmS, which stand for triplet-minus-singlet and doublet-minus-singlet (respectively), will be used for referring to difference spectra representing the loss of ground-state molecules and the gain of the correspondig photoproduct.

Let us note first two severe disadvantages of using the currently available commercial instruments for performing laser flash photolysis experiments.

In the standard, single-beam version of the nanosecond laser photolysis method for single-wavelength kinetic studies, a Q-switched laser is used as the pump source and $\Delta_w(t)$ provides a measure of the concentrations of the absorbing species. The pump and probe beams travel along perpendicular directions in a T-shaped arrangement, with the wide pump beam traveling in the direction of the stem of the "T". The sample is placed at the intersection of the two arms, with the monitoring beam travelling along the cross arm. The monitoring source (a pulsed xenon arc or a flash lamp) is placed on one side of the sample and a photomultiplier tube (PMT) on the opposite side. A monochromator is placed between the sample and the PMT, and an additional monochromator or an optical filter between the sample and the probe source. With this arrangement, $\Delta_w(t)$ can be calculated only if the intensity of the monitoring beam remains constant during the time interval of interest, in which case $I_w(t)$ can be identified with the constant pre-laser value of the PMT signal. When this condition is not fulfilled, a second measurement of the PMT output is made (after blocking the path of the laser beam) to acquire a reference signal, but shot-to-shot variations in the output of the monitoring source vitiate the analysis when the peak value of $\Delta_w(t)$, is small. To overcome this problem, we developed a double-beam version in which an optical fiber and a PMT measure a part of the monitoring beam before it enters the cuvette holder and the output of this PMT is used as the reference signal [26].

A second disadvantage of commercial instruments, associated with the short optical path length (typically 1 cm), will be described by assuming that one is monitoring $T_n \leftarrow T_1$ absorption, which has at its peak value at the end of the laser pulse ($t = 0$); the argument can be easily modified to deal with other photoproducts. It is difficult to achieve a good signal-to-noise ratio if $\Delta_w(0)$ is significantly smaller than 0.1, which implies that even if $\varepsilon_w^\dagger$ is as large as 10^4, one must have $[\mathbf{F}^\dagger]_0 = 10^{-5}$ in order to get $\Delta_w(0) = 0.1$ With $k_1' = 10^3$ s^{-1} and $k_2 = 10^9$ M^{-1} s^{-1}, one finds that $k_2[\mathbf{F}^\dagger]_0 = 10\ k_1'$; that is to say, the decay is dominated by processes 9–11 in Scheme 2, and it is virtually impossible to extract the value of k_1' from the decay curve. One solution to this problem is to abandon the popular right-angle geometry and build an instrument that uses the so-called in-line geometry in which the reference (probe) beam is collinear with the pump beam, but this arrangement is not practicable if one is interested in long path lengths (10 cm or longer). The only practical solution is to use a different technique for following the decay of the triplet population, and our choice is described next.

Though publications reporting E-type delayed fluorescence and phosphorescence from rigid solutions of flavins have appeared in the literature, we have not come across any reports of delayed emission from fluid solutions. However, it seemed worthwhile to us to construct a pulsed-source phosphorimeter and see if delayed emission from an aqueous solution could be observed. The only obstacle to making reliable measurements of the intensity of delayed emission is the occurrence, much more often than not, of an intense burst of prompt emission that drives the PMT into saturation. Mechanical choppers and time-correlated single photon counting are not convenient for lifetimes of 0.1–1 ms. A popular way to overcome detector saturation is to suppress the gain of the PMT for a short interval; an alternative is the use of a specially designed bleeder circuit that delivers a large current for a short period, and allows the PMT to recover normal linear response sufficiently rapidly after its photocathode is flooded by a burst of prompt fluorescence. In a recent publication [27], we have compared the performances of a new electronically gated PMT module and an ungated PMT, and demonstrated that both are suitable for recording delayed emission with a lifetimes longer than a few tens of μs.

3. Experimental Methods

Since the details of our pump-probe arrangement for making time-resolved measurements of photo-induced changes in the absorption spectrum of the sample have been presented in several publications [26, 28–30], only a brief account will be given here.

The excitation source (hereafter called the pump source) was a light pulse of about 6 ns duration; its wavelength can be tuned over the 200-1000 nm interval. The pulse is derived from a single unit containing the following components: a Nd-YAG laser, nonlinear crystals for frequency multiplication by a factor of 2, 3 or 4, an optical parametric oscillator (OPO) pumped by the third harmonic and nonlinear crystals for extending the output of the OPO into the ultraviolet. Two units, both manufactured by EKSPLA, were used: transient absorbance measurements were made with a Model NT 342B-SH-10-WW and delayed fluorescence measurements with a Model NT 342A-SH-10-WW. The construction and performance of the pulsed-source phosphorimeter used for measuring triplet lifetimes have also been presented in a recent publication [27], and repetition here is not necessary.

Differential absorption spectra were recorded using an ungated, fiber-coupled 2048-pixel spectrograph (B&WTEK Model BRC642E), in which each pixel has a width of 14 μm. The spectrograph covers a rather wide spectra range (190–1100 nm) and provides a reciprocal linear dispersion of 0.47 nm per pixel. The probe beam, which traversed at a right-angle to the excitation beam, was delivered by a miniature xenon flash lamp (Hamamatsu L9456-01) with a duration of about 1 μs and a FWHM of about 500 ns. Time-resolved spectra were acquired by varying the delay (denoted by t) between the firing of the pump flash and the triggering of the probe source. The overall time resolution, determined by the duration of the probe flash, is close to 1 μs. When dealing with scattering and/or luminescent samples, a correction for the signal generated by spurious photons (scattered or emitted by the sample into the collection element) is made by recording the output of the spectrograph after firing the pump source alone; this correction is indispensible when one is recording small changes in absorbance. The photo-induced changes in the absorbance of the sample are displayed in the form of a difference spectrum equalling, $\Delta A(\lambda;t) \equiv \tilde{A}(\lambda;t) - A(\lambda)$, where $A(\lambda)$ is the absorbance of the sample before, and $\tilde{A}(\lambda;t)$ the absorbance at time t after the instant at which the pump source is fired.

A suitably modified version of a commercial instrument (Applied Photophysics model LSK-160) was employed for performing single-wavelength kinetic spectroscopy, which entails recording the time course of $\Delta_w(t)$ by using a pulsed xenon lamp as the source of probing light; the modification, which has been described in a recent publication from our group [26], consists in recording simultaneously the shape of the probe pulse by an additional detector.

Time-resolved emission spectra were acquired by means of an electronically gated, 1024-channel spectrograph (Hamamatsu C10029-01), which has a minimum gate width of 10 ns and covers the region 200–860 nm. The time-dependence of the delayed emission (delayed fluorescence and phosphorescence) was investigated with a pulsed-source phosphorimeter whose construction has also been described recently [27]. The sample holder is either a standard 1 cm quartz cell, or, when temperature control is desired, a water-jacketed quartz cell (Starna 63-Q-10) with two polished faces (on opposite sides), the other two frosted. The laser beam enters the jacketted cuvette through a frosted face, and the light emitted at a right angle and passing through a polished face is detected by a time-resolved phosphorimeter comprising a photomuliplier tube (PMT) and a digital oscilloscope, if the time profile of delayed luminescence is to be recorded, or transported by an optical fiber to a gated spectrograph [27] if the spectrum of delayed emission is to be acquired at a certain delay after the laser pulse. Kinetic data were recorded using a commercial gated PMT module (Hamamatsu H11526-20-NN) gated off for a period of 25 μs after the laser pulse. The phosphorimeter is used in a T-shaped configuration during routine measurements, with a concave mirror on one cross arm and the detector (PMT or the input fiber of the spectrograph) on the other. In one set of measurements, the reason for performing which will be explained later, simultaneous measurements of the outputs of the two detectors were made by placing each detector on a cross arm.

The emission spectra reported here have not been corrected for the wavelength sensitivity of the spectrograph. All measurements were performed at 294 K, using freshly prepared solutions which had been deoxygenated by argon bubbling for 30 min prior to photolysis.

4. Results and Analysis

We have examined both lumiflavin and riboflavin, and found no significant difference between their behaviors as regards the processes described above; most of the results presented below refer to lumiflavin. Two hydrophobic carotenoids (astaxanthin and ethyl β-apo-8′-carotenoate) and three hydrophilic carotenoids (crocin, norbixin and astalysine) were studied, but the results shown below refer exclusively to the hydrophilic carotenoids. Some recent spectroscopic data concerning the three hydrophilic carotenoids were recently presented by Chábera et al. [31]. Molar absorption coefficients of the photoproducts of $^{L}\mathbf{F}$ and $^{R}\mathbf{F}$ in aqueous solvents have been determined in our laboratory [30]. We note, for example, that the the molar absorption coefficient at 710 nm for the triplet state of lumiflavin and riboflavin ($\varepsilon^{\dagger}_{710}$) is close to 7×10^4 M^{-1} cm^{-1}.

The absorption spectra of aqueous mixtures containing lumiflavin/crocin and riboflavin/norbixin are displayed in Figures 3 and 4, respectively.

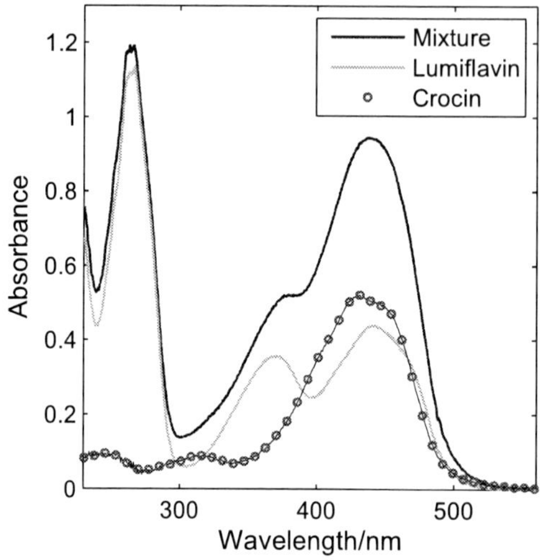

Figure 3. Absorption spectrum of an aqueous mixture containing lumiflavin (33.3 μM) and crocin (4.3 μM), and the absorption spectra of the individual components.

4.1. Delayed Emission: Identification and Characteristics

Setting $k = k_1' + k_q[Q]$ in equation (23), and using $[\mathbf{F}^{\dagger}]_0$ to denote the concentration of flavin triplets at the end of the laser pulse (idealized as a δ-pulse),

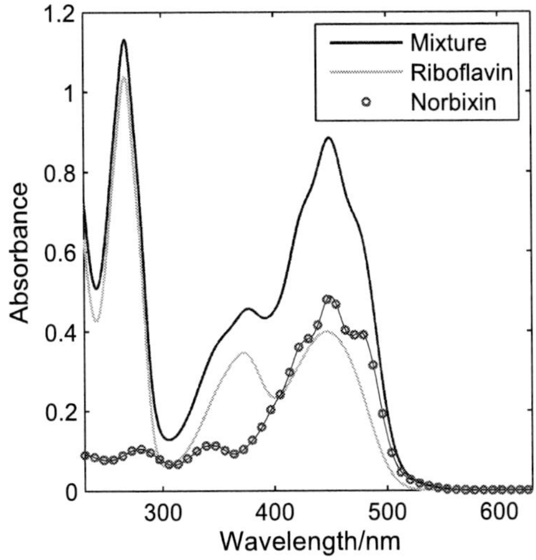

Figure 4. Absorption spectrum of an aqueous mixture containing riboflavin (33.3 μM) and norbixin (4.3 μM), and the absorption spectra of the individual components.

the time dependence of the triplet concentration can be expressed as [21, 32]

$$[\mathbf{F}^{\dagger}] = \frac{k[\mathbf{F}^{\dagger}]_0 \exp(-kt)}{k + k_2[\mathbf{F}^{\dagger}]_0 \{1 - \exp(-kt)\}}. \tag{24}$$

The time $t = 0$ is to be understood as an instant at which essentially all the excited singlets generated by the laser pulse have decayed but the triplets formed through intersystem crossing have suffered negligible depopulation. Since the fluorescence lifetime of flavins is about 5 ns [19, 33], the time origin may be taken as 50 ns.

The rate equation for the generation and decay of $[\mathbf{F}^*]$ on account of the sequence

$$\mathbf{F}^{\dagger} \xrightarrow[\text{process 5}]{k_{\mathrm{TS}}} \mathbf{F}^* \xrightarrow[\text{processes 2,3,4}]{k_{\mathrm{FL}},\, k_{\mathrm{SG}},\, k_{\mathrm{ST}}} \mathbf{F}^0 + h\nu_{\mathrm{F}},\ \mathbf{F}^0,\ \mathbf{F}^{\dagger} \tag{25}$$

can be expressed as

$$\frac{\mathrm{d}[\mathbf{F}^*]}{\mathrm{d}t} = k_{\mathrm{TS}}[\mathbf{F}^{\dagger}] - k_{\mathrm{S1}}[\mathbf{F}]^*. \tag{26}$$

Since we are interested in times which are long compared with the fluorescence lifetime of flavins ($k_{S1}t \gg 1$), we can invoke the steady-state approximation by setting $d\mathbf{F}^*/dt = 0$, which yields

$$[\mathbf{F}^*] = \frac{k_{TS}}{k_{S1}}[\mathbf{F}^\dagger]. \tag{27}$$

Similarly, the rate equation for the formation and disappearance of $[\mathbf{F}^*]$ through the sequence shown below

$$\mathbf{F}^\dagger + \mathbf{F}^\dagger \xrightarrow[\text{process 10}]{k_{2b}} \boxed{\mathbf{F}^*} + \mathbf{F}^0 \xrightarrow[\text{processes 2,3,4}]{k_{FL}, k_{SG}, k_{ST}} \boxed{\mathbf{F}^0 + h\nu_F,\ \mathbf{F}^0,\ \mathbf{F}^\dagger} + \mathbf{F}^0 \tag{28}$$

leads, after the enforcement of the steady-state approximation, to the following expression for the concentration of $\mathbf{F}^*$:

$$[\mathbf{F}^*] = \frac{k_{2a}}{k_{S1}}[\mathbf{F}^\dagger]^2. \tag{29}$$

Let the intensities of phosphorescence, E-type delayed fluorescence and P-type delayed fluorescence be denoted by Π, E and P, respectively. It follows from equation (24) that these intensities can be expressed as

$$\Pi = k_{PH}[\mathbf{F}^\dagger] = \frac{\alpha_1 \Pi_0 \exp(-kt)}{\alpha_1 + \beta_1\{1 - \exp(-kt)\}}, \tag{30}$$

$$E = k_{FL}[\mathbf{F}^*] = \frac{\alpha_2 E_0 \exp(-kt)}{\alpha_2 + \beta_2\{1 - \exp(-kt)\}}, \tag{31}$$

$$P = k_{FL}[\mathbf{F}^*] = \frac{\alpha_3 P_0 \exp(-2kt)}{[\alpha_3 + \beta_3\{1 - \exp(-kt)\}]^2}, \tag{32}$$

where α_i and β_i are constants whose expressions involve $\mathbf{F}_0^\dagger$, the relevant rate constants, and instrumental sensitivity.

If the condition $k \gg k_2[\mathbf{F}^\dagger]_0$ is satisfied, the right-hand side of equation (24) simplifies to $[\mathbf{F}^\dagger] = [\mathbf{F}^\dagger]_0 \exp(-kt)$. If $[\mathbf{F}^\dagger]$ follows exponential decay with a rate constant k, the intensity of E-type delayed fluorescence, being proportional to $[\mathbf{F}^\dagger]$ [34], will follow the same time course, and that of P-type delayed fluorescence, being proportional to $[\mathbf{F}^\dagger]^2$ [35], will also decay exponentially but with with a rate constant equal to $2k$. All lifetimes measured in this study were obtained from data which conformed to exponential decay, which means that the PMT was exposed to either pure E-type or pure P-type delayed fluorescence.

The two cases can be distinguished by investigating the dependence of Y_0, the initial intensity of the delayed emission, on the magnitude of $[\mathbf{F}^\dagger]_0$.

The easiest solution, which requires examining the dependence of Y_0 on the intensity of the laser pulse, could not be implemented since we were not able to adjust the output power of the laser in a controlled and quantifiable manner; changing the wavelength of excitation and/or using a filter has/have been used in this work as a means of varying the initial triplet concentration. Since $[\mathbf{F}^\dagger]_0 = \phi^\dagger[\mathbf{F}^*]_0$, an alternative approach, involving simultaneous kinetic and spectral measurements, was therefore used, involving the investigation of the dependence of Y_0 on $[\mathbf{F}^*]_0$. For a fluorophore with a lifetime of about 5 ns [19, 33], the fluorescence signal obtained with a gate width of 50 ns provides a measure of $\mathcal{F}$, the time-integrated intensity of prompt fluorescence, which is proportional to $[\mathbf{F}^*]_0$. A plot of Y_0 against $\mathcal{F}$, shown in Figure 5, displays the expected linear relation between Y_0 and $\mathcal{F}$ with a negligible intercept on the vertical axis, whence follows the conclusion that only E-type delayed fluorescence entered the PMT in the experiments used for determining the triplet lifetimes.

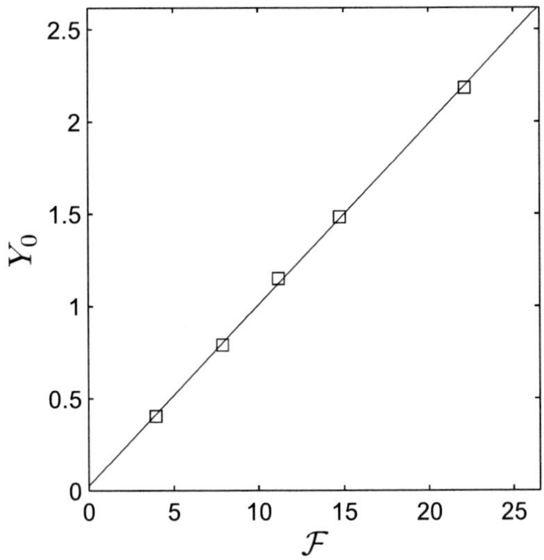

Figure 5. Plot showing the dependence of Y_0 (the initial amplitude of the delayed emission signal) on $\mathcal{F}$ (the integrated intensity of prompt fluorescence).

We pause at this stage to point out that a gated spectrograph, though extremely convenient, is not indispensable for investigating the relationship between Y_0 and $\mathcal{F}$. If an ungated spectrograph is used instead, it will measure $\mathcal{T} = \mathcal{F} + \mathcal{D}$, the integrated intensity of total emission (prompt and delayed), which can be equated, for all practical purposes, to $\mathcal{F}$, since $\mathcal{D}$, the integrated intensity of delayed emission (delayed fluorescence and phosophorescence) is negligible in comparison with $\mathcal{F}$.

The ambiguity can be resolved even if no spectrograph is available. To show how this can be done, we introduce, for later use, three abbreviations, in which A, B_1, B_2, M_0, α and $\tilde{k}$ are constants, to be treated later as adjustable parameters: $\mathcal{S} \equiv A\exp(-\tilde{k}t)$, $\mathcal{B} \equiv B_1\exp(-\tilde{k}t) + B_2\exp(-2\tilde{k}t)$ and $\mathcal{M} \equiv M_0\exp(-\tilde{k}t)/[\tilde{k} + \alpha\{1 - \exp(-\tilde{k}t)\}]$. The functions $\mathcal{S}$, $\mathcal{B}$, and $\mathcal{M}$ will be used for describing (respectively) decays of the following type: single-exponential, bi-exponential, and mixed.

Let us pause to consider the situation (to be called Case I) where $k_2[\mathbf{F}^\dagger]_0$ is non-vanishing but very much smaller than k. In this case, the triplet concentration will appear to follow single exponential decay, and we will have $\Pi = \Pi_0\exp(-kt)$, $E = E_0\exp(-kt)$ and $P = P_0\exp(-2kt)$, and the intensity of delayed emission will, in general, be a bi-exponential, an example of which we have recently provided [27]. We will use the symbol y_1 to denote the PMT output for this case, and the symbol $\models$ for the expression "can be modeled as". When experimental conditions correspond to Case I, $y_1 \models \mathcal{S}$; conversely, the statement $y_1 \models \mathcal{S}$ implies, first, that $k_2[\mathbf{F}^\dagger]_0$ is negligible in comparison with k, and secondly, *either* $\tilde{k} = k$ (the signal is dominated by phosphorescence and/or E-type delayed fluorescence) *or* $\tilde{k} = 2k$ (the signal is dominated by P-type delayed fluorescence). We will now describe how the ambiguity is to be resolved.

Two preliminary statements are needed here by way of explanation: first, if P-type delayed fluorescence is dominant for a particular value of $[\mathbf{F}^\dagger]_0$, the dominance will be accentuated after $[\mathbf{F}^\dagger]_0$ is made larger (other things remaining the same); second, if E-type delayed fluorescence dominates for a particular value of $[\mathbf{F}^\dagger]_0$, it may cease to do so at a higher value of $[\mathbf{F}^\dagger]_0$. Our strategy is to increase the initial triplet concentration to a level at which Case I no longer holds and ask whether y_2, the new PMT output, conforms with a particular model or fails to do so. If $y_2 \models \mathcal{M}$ (Case IIa), or $y_2 \models \mathcal{B}$ (Case IIb), the ambiguity is resolved: P-type delayed fluorescence makes a negligible contribution to y_1 and the best-fit value of the decay constant is to be identified with k rather than $2k$. E-type delayed fluorescence remains dominant in Case IIa, but not in Case IIb. Since these considerations, which can only be applied when the signal-to-noise ratio is large enough to permit discrimination between different models, are sufficient for analyzing our data, we will not discuss other possibilities.

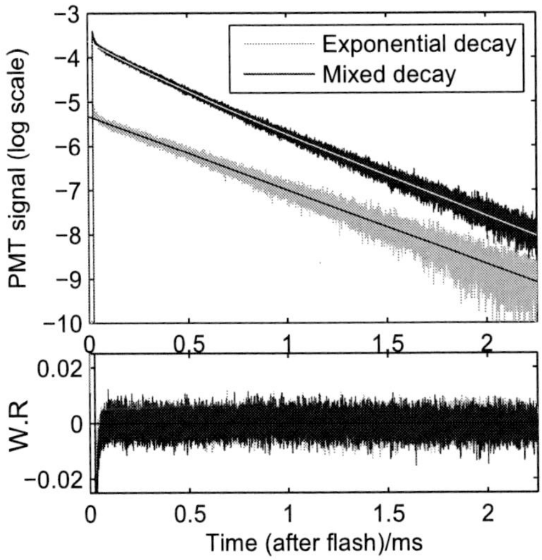

Figure 6. Semi-logarithmic plots of the intensity of delayed emission. The smooth traces show mixed decay predicted by equation (24) (upper curve) and single-exponential decay (lower curve); also shown are the corresponding residuals, weighted according to a Poisson distribution, which are statistically identical. The noisy plots were obtained by recording the delayed emission following laser excitation at 460 nm (upper plot) and 468 nm (lower plot) of a deaerated 2.8 μM solution of lumiflavin in water.

The time profile of delayed emission for a dilute solution of lumiflavin is exhibited in Figure 6, in which the lower and upper traces correspond to y_1 and y_2, respectively; as explained elsewhere [27], the initial, rapidly decaying part of the signal is an artefact resulting from the intense burst of prompt fluorescence. Since we were able to verify that $y_1 \models \mathcal{S}$ and $y_2 \models \mathcal{M}$, we feel justified in concluding that the delayed emission viewed by the PMT is not P-type delayed fluorescence. The good quality of the fit lends confidence to the Stern-Volmer plots, to be presented below, which were all obtained by using PMT outputs that can be modeled as mono-exponential decays. Since the uncertainty in the retrieved value of the decay constant does not exceed 1%, the scatter in our Stern-Volmer plots was so small that the standard deviation of the slope of a linear fit turned out to be negligible. The results reported here are arithmetic means of three separate determinations, and the uncertainty has been taken as 3σ, where σ is the standard deviation of the three values of the slope.

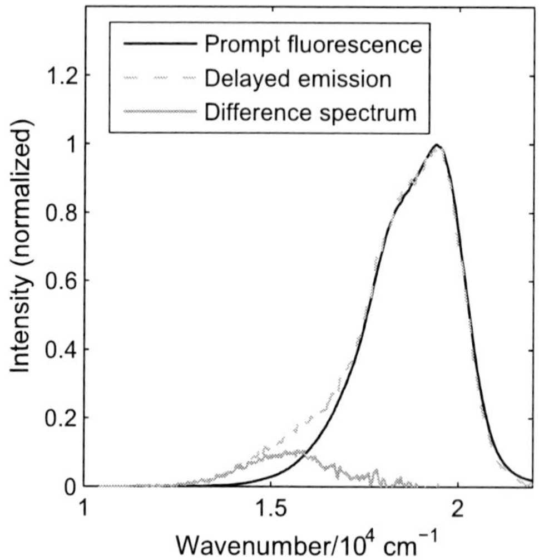

Figure 7. Normalized emission spectra of a 2 μM aqueous solution of lumiflavin excited at 441 nm. The larger relative intensity of the delayed emission at longer wavelengths is attributed to phosphorescence emission whose spectrum was obtained by plotting the difference between the two normalized spectra.

As stated above, the relative intensity of the delayed emission is rather weak, and the light leaving the sample was therefore not filtered before it entered the PMT. In order to make sure that that no spurious emission reached the photocathode, we recorded the uncorrected spectral distribution of the light emitted by the sample within two time windows, to be called narrow and wide, during which the detector is exposed to only prompt and delayed emission, respectively. For the prompt window, the delay time δ was set to 0.5 ns and the gate width ω to 20 ns; for the wide window, $\delta = 1$ μs and $\omega = 2$ ms. Figure 7 displays the two spectra, which have been normalized to the same amplitude at the peak wavelength of the prompt spectrum, and their difference. On comparing the shapes of these spectra with analogous spectra from rigid solutions of lumiflavin [36], it is reasonable to assign our difference spectrum (to be called the red band) to the phosphorescence emitted by the sample, and the delayed spectrum represents, by the same token, a superposition of delayed fluorescence and phosphorescence. Since experiments with riboflavin yielded analogous sresults, we feel justified in ruling out the possibility that the red band originated from an impurity. As further confirmation, we recorded the delayed emission under conditions more favourable to phosphorescence emission by using a 1:1 (v/v) water:glycerol mixture as the solvent and by lowering the temperature of the sample to 278 K; the resulting spectrum is shown in Figure 8.

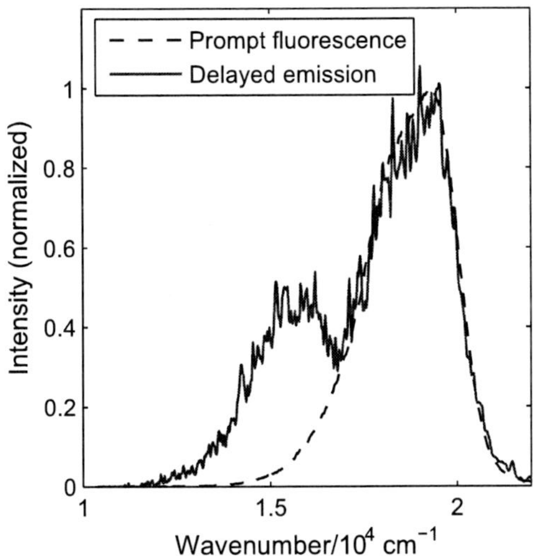

Figure 8. Prompt and delayed emission spectra of a 2.5 μM solution of lumiflavin in a 1:1 (v/v) water:glycerol mixture at 278 K.

The temperature dependence of the initial amplitude of the delayed fluorescence signal over the temperature range 280–310 K is shown in the semi-logarithmic plot appearing in Figure 9. The value of ΔE, the activation energy (for reverse intersystem crossing to the singlet manifold), was found by fitting the data to the Arrhenius relation $\ln Y_0 = A\exp(-\Delta E/k_{\mathrm{B}}T)$, and came out to be $(2.7 \pm 0.2) \times 10^3$ cm^{-1}.

The large activation energy explains why E-type delayed fluorescence is so faint at room temperature, but we still have to account for the fact that P-type delayed fluorescence is still weaker. We have argued earlier [11] that ion pair formation (process 9) dominates other triplet-triplet interactions ($k_{2a} \gg k_{2b}, k_{2c}$); the extremely low intensity of P-type delayed fluorescence becomes intelligible if we postulate that ion recombination, if it occurs, does not lead to the formation

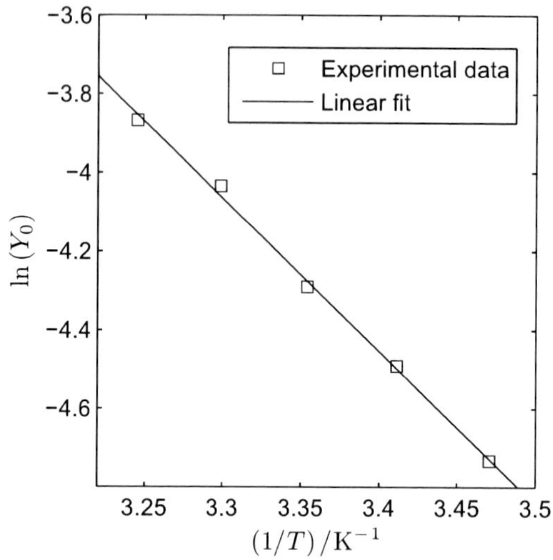

Figure 9. Arrhenius plot showing the temperature-dependence, in the range 288–308 K, of the initial amplitude of the delayed fluorescence signal.

of an singlet excited state. According to this description, most triplet-triplet encounters follow the reactions shown below:

$$\mathbf{F}^{\dagger} + \mathbf{F}^{\dagger} \xrightarrow{k_{2a}} \mathbf{F}^{\bullet +} + \mathbf{F}^{\bullet -} \xrightarrow[+\mathrm{H}^{+}]{} \mathbf{F}^{\bullet +} + \mathbf{FH}^{\bullet}. \quad (33)$$

4.2. Quenching Experiments

We will begin by discussing the quenching of a flavin triplet by a ground-state molecule. A plot of the measured first-order rate constant $k_1' = k_1 + k_3[\mathbf{F}^0]$ against $[\mathbf{F}^0]$, shown in Figure 10, conforms to the expected linear behavior; the intercept on the vertical axis gives $k_1 = (7.30 \pm 0.15) \times 10^2\ \mathrm{s}^{-1}$ and the slope comes out to be $k_3 = (1.1 \pm 0.06) \times 10^8\ \mathrm{M}^{-1}\ \mathrm{s}^{-1}$. One sees, upon comparing our results with those listed in Table 1, that only one previous study [24] reported a value close to ours; however, this result was subsequently retracted and replaced by an upper limit [25]: $k_3 < 2 \times 10^7\ \mathrm{M}^{-1}\ \mathrm{s}^{-1}$. The quality of our data is clearly superior to that of previously reported results, and there is every reason to believe that our investigation has improved on previous attempts to determine k_3.

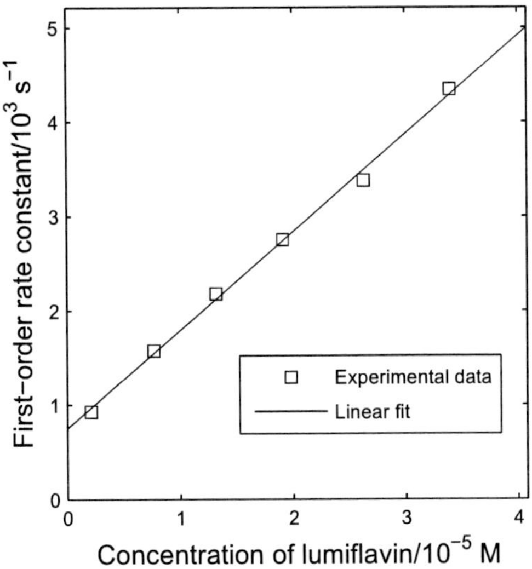

Figure 10. Plots of the measured first-order decay time constants against the concentration of lumiflavin in water.

We move now to the quenching of flavin triplets by two amino acids (cysteine and tryptophan) and the hydrophilic carotenoid crocin. A Stern-Volmer plot for the quenching of lumiflavin triplet by cysteine (in phosphate buffer, pH 6.8) appears in Fig. 11; the slope of the straight line, which can be identified with k_q, came out to be $(1.1 \pm 0.06) \times 10^7$ M^{-1} s^{-1}, in good agreement with that found by laser photolysis [16]. Our values of k_q for quenching by tryptophan, crocin and cysteine in aqueous solutions of lumiflavin came out to be $(2.7 \pm 0.06) \times 10^9$ M^{-1} s^{-1}, $(2.5 \pm 0.06) \times 10^9$ M^{-1} s^{-1} and $(3.9 \pm 0.2) \times 10^7$ M^{-1} s^{-1}, respectively

4.3. Laser Flash Photolysis

It follows from equation (24) that the transient absorbance at a wavelength where only $\mathbf{F}^\dagger$ absorbs can be expressed as

$$\Delta_w = \frac{\Delta_{w,0} \exp(-kt)}{1 + (k_2 C_0/k)\{1 - \exp(-kt)\}}, \tag{34}$$

where $\Delta_{w,0}$ indicates the initial value of Δ_w, $C_0 = \Delta_{w,0}/\hat{\varepsilon}_w^\dagger$, and $\hat{\varepsilon}_w^\dagger = \varepsilon_w^\dagger l$. A special case of some interest is that when $kt \ll 1$, since one can then use the

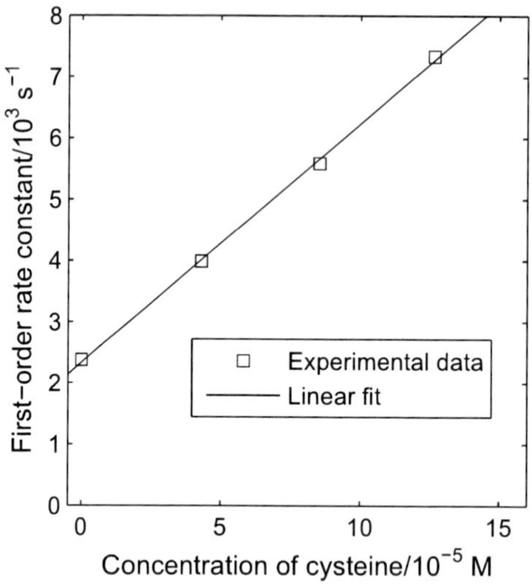

Figure 11. Plots of the measured first-order decay time constants against the concentration of cysteine in anaerobic buffer (pH 6.8).

approximate relations

$$\exp(-kt) = 1 - kt = \frac{1}{1+kt}, \tag{35}$$

which are valid when terms involving $(kt)^n$ can be ignored for $n \geq 2$, to deduce the relation

$$\Delta_w = \frac{\Delta_{w,0}}{1+(k+k_2C_0)t}, \tag{36}$$

which can be further manipulated to get

$$\frac{1}{\Delta_w} = \frac{1}{\Delta_{w,0}} + \left(\frac{k}{\Delta_{w,0}} + \frac{k_2}{\hat{\varepsilon}_w^\dagger}\right)t. \tag{37}$$

We begin by discussing laser-induced changes in quencher-free samples, and display in Figure 12 the time-dependence of Δ_{715} for three solutions of lumiflavin in water (with concentrations 20, 26.7 and 40 μM). Although all three curves are well fitted by equation (24), the retrieved values of the parameters k_1 and k_2 cannot be considered reliable. For example, the values of k_1 (i.e. of k_1' with zero or negligible concentrations of both O_2 and Q) are not consistent with the values found above by using the delayed fluorescence data. One

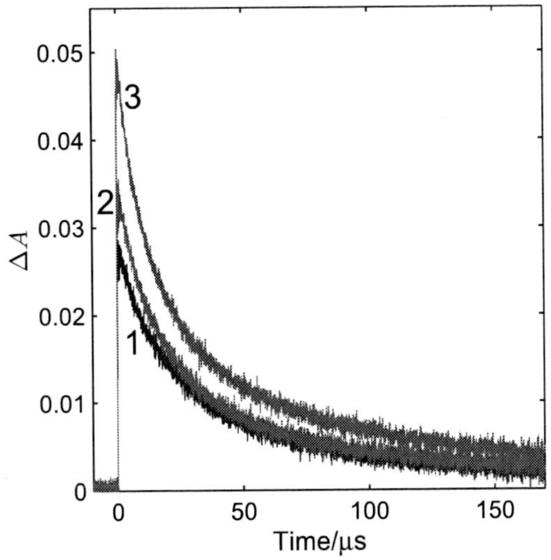

Figure 12. Plots of Δ_{715} against time, obtained by exciting aqueous solutions of lumiflavin at 441 nm; the concentrations are 19.7 μM (curve 1), 26.2 μM (curve 2) and 39.4 μM (curve 3).

source of difficulty is the occurrence of process 9, which generates, after an encounter between two flavin triplets, of $\mathbf{F}^{\bullet+}$ and $\mathbf{F}^{\bullet-}$, which are both paramagnetic and therefore excellent quenchers of triplet states [25]. This means that equation (23), and therefore its solution, stated in equation (24), is inadequate for describing the decay of flavin triplets under conditions which are employed in a typical laser photolysis experiment. Since equation (24) is expected to provide a reasonable description at short times, it is interesting to see if reliable results can be obtained by plotting $1/\Delta_{715}$ against t, and focusing attention on the early part of the decay ($kt \ll 1$); the data for all three solutions fall on a straight line, as shown in Figure 13, and only begin to deviate appreciably from this linearity at $t > 30\mu s$.

Tinkler and coworkers [37] made the interesting observation that an encounter between a 1-nitronaphthalene triplet ($\mathbf{1NN}^{\dagger}$) and a carotenoid may generate not only the carotenoid triplet but also, if the solvent is sufficiently polar, a carotenoid radical cation. We wished to see if competitive electron and energy transfer could be demonstrated in the quenching of $\mathbf{F}^{\dagger}$ by $\mathbf{C}^{0}$; the reactions of interest and the symbols for the two rate constants are shown below:

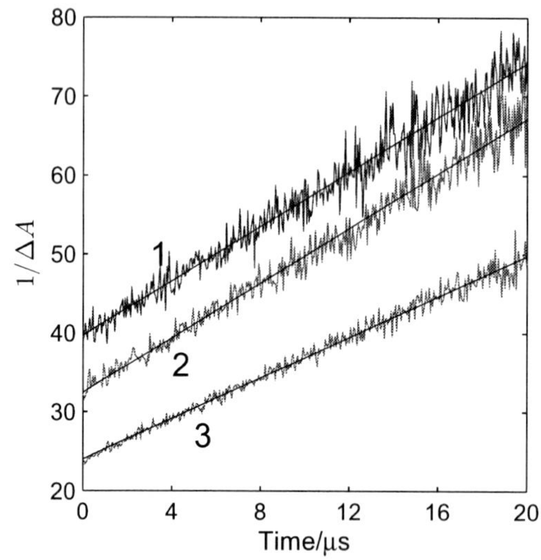

Figure 13. Plots of $1/\Delta_{715}$ against time; other details as in Figure 12.

$$\mathbf{F}^{\dagger} + \mathbf{C}^{0} \longrightarrow \begin{cases} \xrightarrow{k_{4a}} \mathbf{F}^{0} + \mathbf{C}^{\dagger} & (14') \\ \xrightarrow{k_{4b}} \mathbf{F}^{\bullet -} + \mathbf{C}^{\bullet +} & (15') \end{cases}$$

Figure 14 shows Δ@1, the differential absorption spectrum (recorded at a delay of 1 μs) of a mixture of **F** and **C** in water at a delay of 1 μs. We have verified that Δ@1 can be reconstructed as a linear combination of three differential spectra: the TmS spectrum representing the gain/loss of the $\mathbf{F}^{\dagger}/\mathbf{F}^{0}$ pair (to be denoted as Δ_3), the TmS spectrum of the $\mathbf{C}^{\dagger}/\mathbf{C}^{0}$ pair (to be denoted as Δ_3'), the DmS spectrum of the $\mathbf{C}^{\bullet +}/\mathbf{C}^{0}$ pair (to be denoted as Δ_2'), and the TmS spectrum of the $\mathbf{C}^{\dagger}/\mathbf{C}^{0}$ pair (to be denoted as Δ_3'). The three contributions to Δ@1 are plotted in Figure 14. The DmS spectrum of the $\mathbf{FH}^{\bullet}/\mathbf{F}^{0}$ pair (to be denoted as Δ_2) is not noticeable at this delay.

Photoexcitation of a mixture of crocin and lumiflavin leads to the instantaneous formation (on our time scale) of $\mathbf{F}^{\dagger}$ and $\mathbf{C}^{\bullet +}$. At later times the concentration of $\mathbf{F}^{\dagger}$ decreases, processes 14 and 15 come into play, and the concentrations of $\mathbf{C}^{\dagger}$ (a species with a lifetime of a few μs) and $\mathbf{C}^{\bullet +}$ (a much longer-lived species) show an initial rise and a subsequent decline, each at its own rate. Because of its short lifetime ($\tau_c = 1/k_c$), $\mathbf{C}^{\dagger}$ reaches a steady-state after about 10

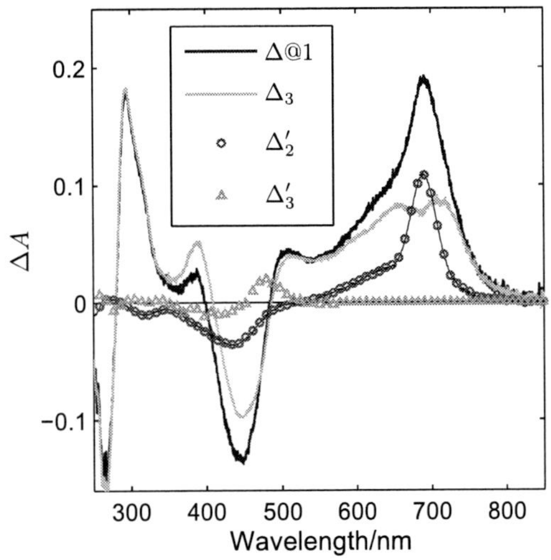

Figure 14. Differential absorption spectrum (Δ@1) of an aqueous solution containing lumiflavin (26.2 μM) and crocin (31.2 μM) recorded at a delay of 1 μs; this spectrum can be reconstructed by the addition of the three other spectra shown here. See text for identification of the component spectra.

μs and follows thereafter the same time dependence as $\mathbf{F}^\dagger$, the sensitizing triplet. Since the concentration of $\mathbf{F}^\dagger$ remains low throughout its existence, and since its absorption spectrum overlaps the spectra of other photoproducts, obtaining a rise and decay curve that fits the relation

$$[\mathbf{C}^\dagger] = \frac{k_{3b}[\mathbf{F}^\dagger]_0}{k_c - k_1}[\exp(-k_1 t) - \exp(-k_c t)] \tag{38}$$

has proved to be an exceedingly difficult task, and we have not yet succeeded in unraveling the kinetics to our satisfaction.

The main objective of a kinetic analysis of the flash photolysis data is to find k_{4a} and k_{4b}, the rate constants for processes 14 and 15, respectively, and this objective can be achieved more simply by relying on other data. We have already found k_q, which can be identified as the sum of k_{4a} and k_{4b}, and our task would be completed if we could find the ratio $r = k_{4a}/k_{4b}$. To see how far one can proceed along this path, we examine a set of time-resolved spectra for a solution containing lumiflavin and crocin (Figure 15), and note the occurrence

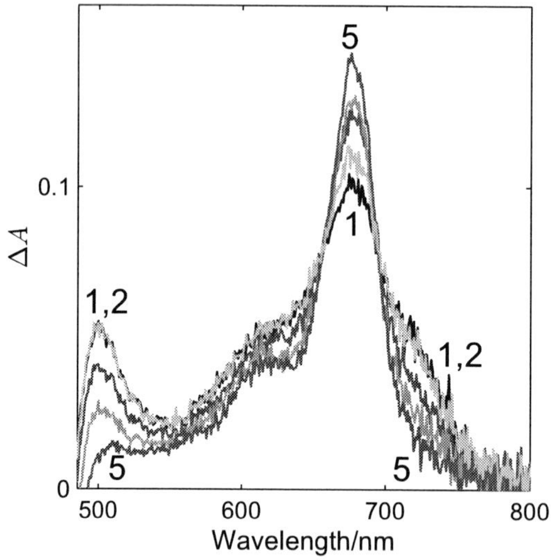

Figure 15. Differential absorption spectra of an aqueous solution containing lumiflavin (31.5 μM) and crocin (7.2 μM) recorded at different delays. Curve 1 (Δ@2) overlaps curve 2 (Δ@5) almost everywhere except in the region between the two isosbestic points; curves 3, 4 and 5 are plots of Δ@15, Δ@25 and Δ@45, respectively. Curve 4 lies above curve 5 at 500 nm, below it at 680 nm.

of two isosbestic points, one at 656 nm and the other at 695 nm. Since only two species, $\mathbf{F}^{\dagger}$ and $\mathbf{C}^{\bullet+}$, contribute to the absorbance at wavelengths longer than 600 nm, we can proceed as follows. Let f denote the probability that an encounter between $\mathbf{F}^{\dagger}$ and $\mathbf{C}$ will give rise to a carotenoid radical cation $\mathbf{C}^{\bullet+}$. It is easy to see that at an isosbestic wavelength (for example, 695 nm), the following relation must hold

$$\varepsilon_{695}^{\dagger\mathrm{L}} = f\varepsilon_{695}^{\oplus\mathrm{C}}. \tag{39}$$

The absorption coefficient $\varepsilon_{w}^{\dagger\mathrm{L}}$ has been determined by us in a previous investigation [30], and the absolute absorption spectrum of crocin was established in this study, leading to the result $\varepsilon_{678}^{\oplus\mathrm{C}} = 2.25 \times \varepsilon_{709}^{\dagger\mathrm{L}} = 2 \times 10^{5}\ \mathrm{M}^{-1}\ \mathrm{cm}^{-1}$. The appropriate value of f can be determined by plotting $S_{w}^{\dagger} \equiv \varepsilon_{w}^{\dagger\mathrm{L}}$ and $S_{w}^{\oplus} \equiv f\varepsilon_{w}^{\oplus\mathrm{C}}$ against wavelength and choosing f so that the condition $S_{695}^{\oplus} = S_{695}^{\dagger}$ is satisfied. If the assumptions underlying our analysis hold, it follows that the condition $S_{656}^{\oplus} = S_{656}^{\dagger}$ should also be satisfied, and this was indeed found to be true. The value of f turned out to be 0.07.

Although we have not been able to find the values of $r = k_{4a}/k_{4b}$, we have reasons to believe that $r < 1$ in water, a highly polar solvent in which electron transfer is expected to dominate over energy transfer. In other words, we rule out the possibility that the probability of energy transfer during an encounter is $1 - f = 0.93$. In order to account for the low yield of ion-pair formation, we are postulate that the second step in the reaction sequence

$$\mathbf{F}^{\dagger} + \mathbf{C}^{0} \xrightarrow{k_{4b}} \mathbf{F}^{\bullet-} + \mathbf{C}^{\bullet+} \longrightarrow \mathbf{F}^{0} + \mathbf{F}^{0}, \tag{40}$$

known as geminate recombination, is far more efficient than the escape of the two radical ions into the bulk solution.

To demonstrate the influence of solvent polarity, we have photolyzed a mixture of lumiflavin and crocin in acetonitrile. A set of three time-resolved spectra is presented in Figure 16. The contributions made by $\mathbf{F}^{\dagger}$ and $\mathbf{F}^{\bullet-}$ have been subtracted from the spectra plotted in Figure 16 to allow a direct comparison of these spectra with those in Figure 14. The occurrence of process 15, leading to the formation of $\mathbf{C}^{\bullet+}$, is evident from the appearance of the 680 nm peak in each set of spectra; the peaks at 482 nm in Figure 16 can be assigned to $\mathbf{C}^{\dagger}$, formed as a result of process 14. One sees that, in keeping with the observations of Tinkler and coauthors [37], electron transfer becomes less probable when water is replaced by the less polar solvent acetonitrile.

Results similar to those reported above were obtained when we used norbixin or astalysine instead of crocin, which indicates that all hydrophilic carotenoids would be good quenchers of flavin triplets, and that quenching would occur through simultaneous operation of processes 15 and 14. We will not present the results pertaining to astalysine, since it is not readily available to other experimenters. A set of time-resolved spectra for an aqueous mixture of lumiflavin and norbixin is presented in Figure 17; the contributions made by $\mathbf{F}^{\dagger}$ and $\mathbf{FH}^{\bullet}$ have been subtracted from the spectra in order to isolate the DmS spectrum of the $\mathbf{C}^{\bullet+}/\mathbf{C}^{0}$ pair.

The concentration of $\mathbf{C}^{\bullet+}$ builds up over a few tens of μs, as shown in Figure 18, a plot of Δ_{780} for a riboflavin-norbixin mixture in aqueous solution.

A summary of our principal results is presented in Table 2. Since most of our work was carried out by using lumiflavin, we have not included the data obtained by using riboflavin. As explained earlier, the value of k_1 depends critically on the solvent purity and the concentration of residual dissolved oxygen.

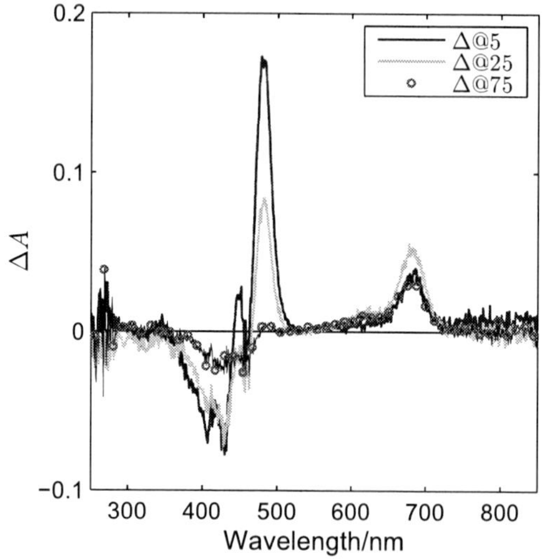

Figure 16. Time-resolved differential absorption spectra obtained at different delays after photolyzing a mixture containing lumiflavin (60 μM) and crocin (15 μM) in acetonitrile; as explained in the test, transient spectra belonging to riboflavin photoproducts have been subtracted.

Table 2. Rate constants determined in this study of lumiflavine by using E-type delayed fluorescence.

Molecule	Parameter	Value
Lumiflavin	k_1	7×10^2 s^{-1}
	k_2	1×10^{10} M^{-1} s^{-1}
	k_3	1.1×10^8 M^{-1} s^{-1}
Tryptophan	k_q	2.7×10^9 M^{-1} s^{-1}
Cysteine (pH 6.8)	k_q	3.9×10^7 M^{-1} s^{-1}
Crocin	k_q	2.5×10^9 M^{-1} s^{-1}

5. Concluding Remarks

As stated in the introduction, this study was designed to assess the feasibility of using hydrophilic carotenoids to protect riboflavin-containing beverages

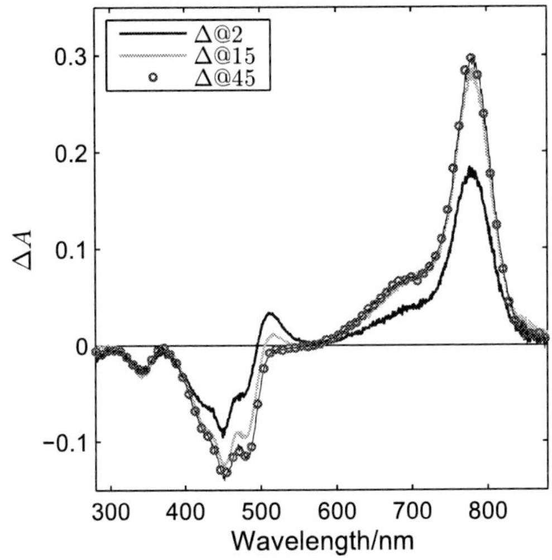

Figure 17. Time-resolved differential absorption spectra obtained at different delays after photolyzing a mixture containing riboflavin (33.3 μM) and norbixin (4.3 μM) in water; as explained in the test, transient spectra belonging to riboflavin photoproducts have been subtracted.

from photodamage. The fact that the amino acids tyrosine and tryptophan are slightly better quenchers than crocin is of little practical significance, since the probability of quenching is determined by the product of k_q and the quencher concentration. Both crocin and norbixin have the advantage that they absorb in the visible region (see Figures 3 and 4) and would therefore also act as inner filters, allowing less light to be absorbed by riboflavin. Leaving monetary considerations aside, the selection of a suitable substrate would entail answering the question: "Which substrate radical cation ($\mathbf{S}^{\bullet+}$), a paramagnetic species, will cause least damage or confer the greatest benefit?" Challenging the notion that free radicals always play a pernicious role in biology, Williams has stressed the need for weighing the risks against the benefits [38]. One advantage, worth mentioning in the context of photoprotection, is the ability of paramagnetic species to act as efficient overall quenchers of electronically excited states [39].

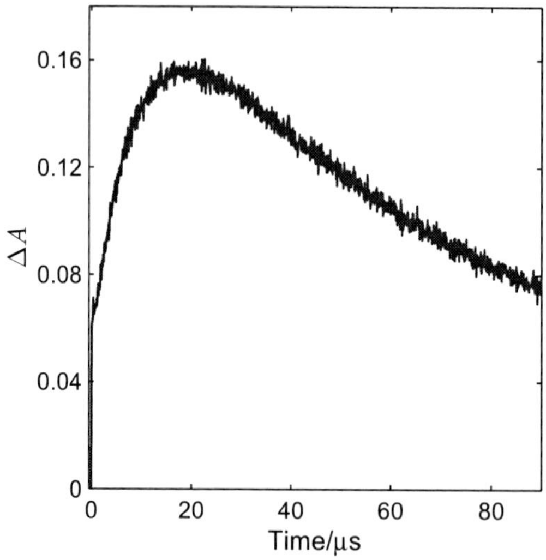

Figure 18. Plot of Δ_{780} against time obtained by photolyzing an aqueous mixture containing riboflavin (33.3 μM) and norbixin (4.3 μM).

Acknowledgments

We are grateful to Vassilia Partali and Hans-Richard Sliwka for their interest in the work and for supplying purified carotenoid samples, and to Robert E. Dale for his critical reading of the manuscript. This work was supported by a grant from the Research Council of Norway (Project No. 191102).

References

[1] Mattevi, A. *Trends Biochem. Sci.* 2006, 31, 276 – 283

[2] van der Horst, M. A.; Hellingwerf, K. J. *Acc. Chem. Res.* 2004, 37, 13–20

[3] Ball, G. F. M. *Vitamins In Foods: Analysis, Bioavailability, and Stability*; CRC Press: Boca Raton, FL, 2005

[4] Heelis, P. F. *Chem. Soc. Rev.* 1982, 11, 15–39

[5] Möglich, A.; Yang, X.; Ayers, R. A.; Moffat, K. *Annu. Rev. Plant Biol.* 2010, *61*, 335–364

[6] Chaves, I.; Pokorny, R.; Byrdin, M.; Hoang, N.; Ritz, T.; Brettel, K.; Essen, L.-O.; van der Horst, G. T. J.; Batschauer, A.; Ahmad, M. *Annu. Rev. Plant Biol.* 2011, 62, 335–364

[7] Christie, J. M. *Annu. Rev. Plant Biol.* 2007, 58, 21–45

[8] Herrou, J.; Crosson, S. *Nature Rev. Microbiol.* 2011, 9, 713–723

[9] Hader, D. P., Jori, G., Eds. *Flavins: Photochemistry and Photobiology*; The Royal Society of Chemistry, 2006

[10] Tsentalovich, Y. P.; Lopez, J.; Hore, P.; Sagdeev, R. *Spectrochim. Acta A* 2002, 58, 2043–2050

[11] Melø, T. B.; Ionescu, M. A.; Haggquist, G. W.; Naqvi, K. R. *Spectrochim. Acta A* 1999, 55, 2299–2307

[12] Cardoso, D. R.; Franco, D. W.; Olsen, K.; Andersen, M. L.; Skibsted, L. *J. Agric. Food Chem.* 2004, 52, 6602?–6606

[13] Goldsmith, M. R.; Rogers, P. J.; Cabral, N. M.; Ghiggino, K. P.; Roddick, F. A. *J. Am. Soc. Brew. Chem.* 1980, 63, 177?–184

[14] Cardoso, D. R.; Olsen, K.; Møller, J. K. S.; Skibsted, L. H. *J. Agric. Food Chem.* 2006, 54, 5630–?5636

[15] Bradley, D. G.; Kim, H. J.; Min, D. B. *J. Agric. Food Chem.* 2006, 54, 6016–?6020

[16] Cardoso, D. R.; Olsen, K.; Skibsted, L. H. *J. Agric. Food Chem.* 2007, 55, 6285–6291

[17] Montenegro, M. A.; Nunes, I. L.; Mercadante, A. Z.; Borsarelli, C. D. *J. Agric. Food Chem.* 2007, 55, 323?–329

[18] Huvaere, K.; Skibsted, L. H. *J. Agric. Food Chem.* 2006, 131, 8049?–8060

[19] Cardoso, D. R.; Surachio, R. S.; Santos, W. G.; de Mello, H. H.; Skibsted, L. H. *J. Agric. Food Chem.* 2013, 61, 2268–2275

[20] Vaish, S. P.; Tollin, G. *J. Bioenerg. Biomembranes* 1970, 1, 181–192

[21] Grellmann, K.; Scholz, H.-G. *Chem. Phys. Lett.* 1979, 62, 64–71

[22] Yoshimura, A.; Kato, S. *Bull. Chem. Soc. Japan* 1985, 58, 1556–1559

[23] Naman, S.; Tegnér, L. *Photochem. Photobiol.* 1986, 43, 331–333

[24] Fritz, B. J.; Matsui, K.; Kasai, S.; Yoshimura, A. *Photohem. Photobiol.* 1987, 45, 539?–541

[25] Yoshimura, A.; Ohno, T. *Photochem. Photobiol.* 1991, 53, 1754?–179

[26] Li, H.; van 't Hag, L.; Yousef, Y. A.; Melø, T. B.; Naqvi, K. R. *Photochem. Photobiol. Sci.* 2013, 12, 404–406

[27] Persvik, Ø.; Melø, T. B.; Naqvi, K. R. *Photochem. Photobiol. Sci.* 2013, 12, 1110–1113

[28] Naqvi, K. R.; Li, H.; Melø, T. B.; Webster, R. D. *J. Phys. Chem. A* 2010, 114, 10795–10802

[29] Zhang, Y.; Yousef, Y. A.; Li, H.; Melø, T. B.; Naqvi, K. R. *J. Phys. Chem. A* 2011, 115, 8242–8247

[30] Li, H.; Melø?, T.; Naqvi, K. R. *J. Photochem. Photobiol. B: Biology* 2012, 106, 34 – 39

[31] Chábera, P.; Fuciman, M.; Naqvi, K. R.; Pólivka, T. *Chem. Phys.* 2010, 373, 56–64

[32] Steel, C.; Naqvi, K. R. *J. Phys. Chem.* 1991, 95, 10713–10718

[33] Gordon-Walker, A.; Penzer, G. R.; Radda, G. K. *Eur. J. Biochem.* 1970, 13, 313–321

[34] Parker, C. A.; Hatchard, C. G. *Trans. Faraday Soc.* 1961, 57, 1894–1904

[35] Parker, C. A.; Hatchard, C. G. *Trans. Faraday Soc.* 1963, 59, 284–295

[36] Penzkofer, A. *Chem. Phys.* 2012, 400, 142–153

[37] Tinkler, J. H.; Tavender, S. M.; Parker, A. W.; McGarvey, D. J.; Mulroy, L.; Truscott, T. G. *J. Am. Chem. Soc.* 1996, 118, 1756?–1761

[38] Williams, R. J. P. *Phil. Trans. Roy. Soc. London B* 1985, 311, 593–603

[39] Naqvi, K. R.; Melø T. B.; Sliwka, H.-R.; Mohamad, S. B. B.; Partali, V. *Photochem. Photobiol. Sci.* 2003, 2, 381?–385

INDEX

E

F

G

H

S

T

U

V

W

X

Y